MEMORIAL TABLET TO BLACK WATCH
IN DUNKELD CATHEDRAL.

From a Photo by J. Valentine & Sons

THE

HIGHLAND BRIGADE:

ITS BATTLES AND ITS HEROES.

ILLUSTRATED.

BY

JAMES CROMB,

AUTHOR OF "THE HIGHLANDS AND HIGHLANDERS OF SCOTLAND";
"WORKING AND LIVING, AND OTHER ESSAYS"; &C.

" We'll ha'e nane but Hieland Bonnets here!"

The Naval & Military Press Ltd

in association with

The National Army Museum, London

Published jointly by

The Naval & Military Press Ltd
Unit 10 Ridgewood Industrial Park,
Uckfield, East Sussex,
TN22 5QE England

Tel: +44 (0) 1825 749494
Fax: +44 (0) 1825 765701

www.naval–military-press.com
www.military-genealogy.com
www.militarymaproom.com

and

The National Army Museum, London
www.national–army–museum.ac.uk

To

LIEUT.-GEN. SIR ARCHIBALD ALISON, BART.,

K. C. B., &c., &c.,

THE PERSONAL FRIEND OF LORD CLYDE,

HIS SUCCESSOR IN THE LEADERSHIP OF THE HIGHLAND
BRIGADE,

AND THE INHERITOR OF HIS SWORD OF HONOUR ;

A BRAVE SOLDIER AND ACCOMPLISHED
GENTLEMAN ;

This Volume

IS,

BY SPECIAL PERMISSION,

RESPECTFULLY DEDICATED

BY

THE AUTHOR.

SIR COLIN'S "THIN RED LINE."

CONTENTS.

Page

INTRODUCTION, ... 9

THE CRIMEA.

Chapter
 I. The Battle of the Alma, ... 15
 II. Do. do. *(continued)*, . . . 28
 III. The Battle of Balaclava, ... 34
 IV. Life in the Trenches, ... 43
 V. Before the Indian Mutiny—The 78th in Persia, . 57

THE INDIAN MUTINY.

 VI. The Outbreak, ... 67
 VII. Advance of Havelock's Column to Cawnpore, . 80
 VIII. Battle of Cawnpore, and Scene in the City, . . 91
 IX. The First and Second Advances on Lucknow, . 103
 X. The Third Advance on Lucknow, . . . 112
 XI. First Relief of Lucknow, ... 123
 XII. After the First Relief of Lucknow, . . . 134
 XIII. The 93rd at Lucknow, ... 147
 XIV. Capture of the Secundra-Bagh, ... 157
 XV. The Second Relief of Lucknow, ... 168
 XVI. Windham's Peril at Cawnpore—Sir Colin's Dash
   to his Relief, ... 180
 XVII. The Third Attack on Lucknow, ... 191
 XVIII. The Capture of Lucknow, ... 199
 XIX. The Campaign in Rohilcund, ... 210
 XX. The Battle of Bareilly, ... 222
 XXI. Close of the 42nd's share in the Campaign, . . 230
 XXII. Return Home of the 93rd, ... 238

RECENT SERVICES OF HIGHLAND REGIMENTS.

Chapter		Page
XXIII.	The 42nd in Ashantee, ...	245
XXIV.	The Afghan Campaigns, . . . - .	263
XXV.	Do. do. (continued), . . .	273
XXVI.	Do. do. do., . . .	280
XXVII.	The 92nd on Majuba Hill, ...	288
XXVIII.	The Highland Brigade at Tel-el-Kebir, . .	297
XXIX.	The Highland Regiments in the Soudan, . .	307
APPENDIX,	. . . .	313

LIST OF ILLUSTRATIONS.

Page

Frontispiece—Memorial to 42nd Highlanders at Dunkeld.

	Page
Sir Colin Campbell, ...	20
General Sir Duncan Alexander Cameron, K.C.B., . .	23
Sketch showing the Highland Brigade's Advance at the Alma,	26
General Sir John Douglas, G.C.B., ...	29
Sketch Map showing the Theatre of War in the Crimea, .	33
Monument in Memory of the Officers and Men of the 79th Highlanders who died in Bulgaria and the Crimea, in the Dean Cemetery, Edinburgh, ...	54
General Outram, ...	59
Sketch Map showing Main Theatre of Indian Mutiny, . .	78
General Havelock, ...	81
Memorial over Well at Cawnpore, ...	100
Mausoleum at Well of Cawnpore, ...	101
General Sir J. E. W. Inglis, ...	114
Sir H. W. Stisted, K.C.B., ...	116
Sketch Plan showing Route of Havelock's Advance to the Residency at Lucknow, ...	131
Monument to the 78th on Castle Esplanade, Edinburgh, .	143

Page

Centre-Piece for Officers' Mess of 78th, presented by the
    Counties of Ross and Cromarty, ... 144

Brigadier-General Adrian Hope, ... 149

Ground Plan of Secundra-Bagh, ... 159

Lieutenant-General F. W. Traill-Burroughs, C.B., .   .   . 166

Bird's-Eye View of Lucknow from River Goomtee,   .   . 171

Sergeant John Paton, V.C., ... 174

East View of Residency at Lucknow—After the Evacuation, . 178

Lieutenant-Colonel Leith Hay, ... 201

Major-General William M'Bean, V.C., ... 205

Major Simpson, V.C., ... 216

Sergeant Alexander Thompson, V.C., ... 218

Flags carried by the 93rd Highlanders through the Indian
    Mutiny War, ... 241

General Lord Wolseley, G.C.B., K.C.M.G., &c.,   .   . 247

General Sir John M'Leod, K.C.B., ... 257

General Sir Frederick Roberts, V.C., ... 265

Colours of the 72nd, Duke of Albany's Own Highlanders,   . 268

Lieutenant-Colonel Brownlow, C.B., 72nd Highlanders,   . 284

General Sir G. P. Colley, ... 294

Granite Monument to Gordon Highlanders, in Duthie Park,
    Aberdeen, ... 296

Lieutenant-General Sir Archibald Alison, K.C.B., .   .   . 299

PREFATORY NOTE TO THIRD EDITION.

THE rapidity with which the two previous editions of "The Highland Brigade" have been bought up has been to me a gratifying surprise. The success of the volume I attribute to the strong patriotic feeling of my countrymen, and to the delight and pride with which they regard the heroic achievements of the tartan-clad regiments. But I am pleased that by gathering the materials together I have, to paraphrase the language of Sir Colin Campbell's famous address, made it easier for my readers, "for their children, and for their children's children, to repeat the tale to other generations." For "our native land will never forget the name of the Highland Brigade." The volume is not to be regarded as a complete history of the services of the Highland regiments during the period with which it deals, but only as a narrative of outstanding events; accurate, I trust, so far as it goes, but not by any means exhausting the subject. Especially is this to be understood in regard to that portion of the volume treating of events from Ashantee to the Soudan. To General Sir Archibald Alison, General Sir H. M. Havelock-Allan, General Sir John Douglas, General F. W. Traill-Burroughs, Colonel Leith-Hay, Colonel Cooper (4th Royal Irish), Adjutant Gordon (1st Gordon Highlanders), and others of humbler rank, I am much indebted for assistance of various kinds, and return them my best thanks. A number of statements and corrections, some of them of considerable interest, have reached me since the publication of the last edition, and for these, as well as for brief notices of the career of two Highland soldiers of reputation—Colonel Duncan Macpherson of Cluny, and General Herbert Macpherson, Commander-in-Chief of the Army of Madras—who have died in the interval, the reader is referred to the Appendices attached to the present volume.

J. C.

DUNDEE, *December* 1886.

INTRODUCTION.

SCOTSMEN all the world over take a patriotic interest in the achievements of the Highland Brigade. They look upon the tartan-clad warriors with peculiar pride, regarding them as approaching more nearly to their ideal of the invincible than any others of the gallant corps of which the British Army is composed. In the opinion of the Scot the honours of the Highlanders shed lustre upon himself; with their prowess he regards himself as in some remote way identified; and he carries his head more proudly as the oft-told story of their victories is spoken in his hearing.

This pride with which the Highlanders are regarded is justifiable. Without disparagement to any of the other brave corps which go to swell the strength of Her Majesty's Army, and whose nobly won honours form an imperishable record, we affirm that each Highland

B

Regiment is entitled to claim the distinction of being a " crack corps." Since the first encounter of the 42nd with the French at St Antoine, the Highlanders have had a heavy share of the fighting in nearly all the campaigns in which the British have been engaged, and have so distinguished themselves that a Highland Brigade is now looked upon as an indispensable part of the composition of every considerable fighting force sent from Britain. They have won the confidence of every General under whom they have served, and have ever been equal to the most trying tasks.

The Regiments at the time of their formation were drawn from a warlike race—much occupied, and highly accomplished, in the art of fighting. The Scottish Highlanders were, in the early part of the 18th century, to a man trained to the use of arms, and it was a wise judgment which suggested the Highlands as a recruiting ground. In all the Sovereign's wide realms could be found no such chivalrous, true-hearted, brave-souled men; nor could they be equalled in those physical qualities which were so much demanded in the harassing system under which war was at the time conducted. Those were the days in which the lithe limb and strong arm were of more importance than they are in the campaigning of to-day, when, with certain powers of endurance understood, science has reduced warfare almost to the dead level of a problem in mathematics.

It was in these qualities of limb that the Highlanders
excelled. They were strong and muscular, accustomed to violent exercises and fatiguing marches.
Their country, with its darksome passes and rugged
heights, its treacherous moors and plunging torrents,
was to a stranger wild and forbidding. But to them it
was a rough training-ground, calculated to bring forth
all that was robust and manly in their frame. That such
a recruiting field should ever have been exhausted—
as practically exhausted it has become—is a thought
which almost tempts the pen into a channel of bitter
reflection. It was of such men that the first Highland
Regiments were composed—true Highlanders every
officer and man. At Fontenoy not a soldier in the
42nd had been born south of the Grampians. At first
they were watched with coldly critical eyes; they were
regarded as semi-barbarians, who could never be subjected to the discipline imposed upon ordinary troops,
as men whose presence might indeed have a prejudicial
effect upon those with whom they were associated.
But they quickly disarmed all such ignorant doubts.
In the barrack-room their conduct was irreproachable;
in the field their heroism was a theme of wonder.
Whatever their post, that post was secure; whatever
their duty, that duty was performed. The greater
the danger opposed to them the greater the honour
achieved. Their charging cheer was the shout of

victory—or, if victory was impossible, it was still the triumphant cry of men who elected to die—their face to the foe, and their latest effort a death grapple.

Since those days the composition of the Highland Regiments has changed—and changed, it is alleged, to an extent which threatens to utterly destroy their distinctive character and serviceableness in the field. The allegation, coming from men who are equally acquainted with the genuine soldierly qualities of true Highlanders and with the rather different qualities of the Cockney recruits who are sometimes drafted into the Highland Regiments, must, we fear, be accepted with less doubt than is agreeable. Yet there is a mighty strength-giving power in the traditions of a crack regiment, and in the associations which cluster round the old flag, which must never be dishonoured. Sentiment will not alone gain victories; but sentiment will sometimes sustain the faltering heart, and give vigour to the nerveless frame. On every recruit who joins a Highland Regiment is thrown the honour of the corps—a charge so precious that none but the veriest poltroon could prove unfaithful to the duty. As yet no sign of the Highlanders losing their prestige or proving unworthy of their traditions has been exhibited. In their most recent engagements they have behaved with that valour for which they have ever been distinguished, and their latest Brigadier—Sir

Archibald Alison—has testified to the pride with which at Tel-el-Kebir he led his men to victory.

The history of the Highland Regiments naturally divides itself into two periods—that preceding and that following Waterloo. Preceding Waterloo the composition of the Regiments was chiefly Highland, although latterly with a strong strain of Lowland blood. After Waterloo the Lowlanders may be said to have preponderated—although the preference for a Highland Regiment of Highlandmen joining the army for long secured a strong Highland element in the whole of the corps. Which of the periods is the more interesting depends much upon the point of view from which the subject is regarded. For ourselves, much as we admire the glorious record of the American, the Southern Indian, the Egyptian, and the Peninsular campaigns, we dwell with equal love on a period which is still well within the memory of men who are living. We have, therefore, selected as the principal subject matter of this volume two campaigns in which Highlanders bore a heavy share of the "storm and stress" of war—first, in the Crimea; and secondly, in India. Both were of the magnitude of great wars; but we also include, very briefly, the story of Highland valour in Afghanistan, Africa, Egypt, and the Soudan. These wars were more desultory in their nature, but the Highlanders' services are not less worthy of being recorded. The two chief

campaigns named cover the whole period during which
the Highland Brigade served under that most popular
of leaders, Sir Colin Campbell, afterwards Lord Clyde,
who led the Highlanders, first as Brigadier in the Crimea,
and afterwards as Commander-in-Chief of the Forces
employed to quell the Indian Mutiny. In the later
campaigns, the Highlanders have been brigaded twice
under Sir Archibald Alison, also a Scotsman, and, like
Sir Colin, a native of St Mungo. Sir Archibald, a
personal friend of Lord Clyde, and the proud inheritor of his sword of honour, led the Black Watch
through the dark African jungle to Coomassie; and
scaled with the tartan-clad heroes the trenches of Tel-
el-Kebir. Under Lord Clyde, the lads of the feather
bonnets won a degree of fame which will survive for
generations in the cities, glens, and homesteads of
Scotland; under Sir Archibald, the traditions of the
glorious past have been well maintained; for the trying
night march, and the desperate daybreak assault, were
as fine an example of soldiering as anything in the
history of Highland military achievement.

THE HIGHLAND BRIGADE.

Chapter I.

FTER Waterloo the first serious war in which the British Army engaged was the Crimean, in 1854. In various parts of the world there had been in the interval fighting more or less severe; but the peace of Europe had been practically unbroken, and no engagements had occurred in which forces of equal magnitude were employed. Britain, France, and Turkey were allied in the operations against Russia. Lord Raglan—the Lord Fitzroy Somerset who had served the Great Duke as Military Secretary in the Peninsular days—was in command of the British force, which included as a Highland Brigade the 42nd, 79th, and 93rd Regiments. The allied force landed at Old Fort, Kalamita Bay, on the Black Sea, and marched on the 19th of September to the Bulganak river, in which direction the Russian Army was expected to be found in force. The march was hot and toilsome, and was a severe strain upon the large numbers of unhardened men in the ranks. Many, too, were weak from sickness, for the voyage had proved trying to the health of the troops. Ever and again as they moved forward men were compelled to fall out, unable to go on, and the first sense of danger was realised, as there appeared imminent risk of these unfortunate fellows falling into the hands of the alert Cossack horsemen, who could be seen hovering on the distant flank. In all the divisions of the army the suffering from thirst and fatigue was very great, and when the men came in sight of the fresh, gladdening waters of the Bulganak they broke from their ranks and rushed forward, as Kinglake says, to "plunge their lips deep in the cool, turbid, grateful stream."

Here the Highlanders, for the first time, discovered that they were in the hands of a stricter disciplinarian than were most of the other brigades. Their Brigadier was Sir Colin Campbell, who had just assumed the command. As an experienced soldier, he knew the high character of the men under his control ; and they knew him as an officer who had made his way by sheer ability to his present position. But they had never campaigned together till now, and dreamed nothing of the deathless fame which they should together win ere to-morrow's sun had set. With stern decision, while others revelled in the cool flood, he kept them to their ranks, and when near the stream halted them in perfect order—determined that not even "the rage of thirst should loosen the discipline of his grand old Highland Regiments." But he had a keen eye to their interests, and when the men saw he had arranged that they should drink and bathe in order and comfort, they felt and said they had been the gainers. And so between commander and men was forged the first link of mutual esteem and admiration.

In the morning it was found that the Russians occupied a strong position on the left bank of the river Alma, a narrow fordable stream a few miles forward. They were strong in numbers and well supported by horse and artillery, counting in all 40,000 men and 106 guns. It was agreed to drive them from the position. The order of battle placed the

French on the right, next the seashore, along which the
French fleet and one English vessel slowly crept, reconnoitring,
and preparing to shell what could be reached of the enemy's
lines. The British force was on the left, marching towards
the Alma in a line parallel with it. On the extreme left
were the Highland Regiments. In this form the armies
advanced slowly in the hot sun. The river ran through
some vineyards, the opposite bank on which the Russians
were posted was sheer and steep, and earthworks and a
redoubt had been thrown up. The attack was not one to be
undertaken lightly, or without careful preparation. Defeat
to the Allies meant irretrievable disaster, and defeat to the
Muscovites was expected to be equally fatal. In truth, this
first meeting of the Eastern and Western Powers of Europe
was felt by both sides to be a momentous occasion; and
although the Russians fully expected to drive the invaders
into the sea, they hesitated about engaging in the dire
attempt. At length the opposing forces were practically face
to face; a brief lull occurred, during which Sir Colin
observed it would "be a good time for the men to get loose
half their cartridges." "When," says Kinglake, "the command travelled on along the ranks of the Highlanders it lit
up the faces of the men one after another, assuring them that
now at length, and after long expectance, they would indeed
go into action. They began obeying the order, and with
beaming joy, for they came of a warlike race; yet not
without emotions of a graver kind; they were young soldiers,
new to battle."

Yet there was an interval of trial to endure. The whole
force had descended into the valley in all the pomp of its
warlike splendour, and, while the left halted and stood still,
the sound of heavy firing far away on the right proclaimed
that the battle had opened.

The conflict had been begun by the French far away on
the right advancing against the Russian position, which,
however, they greatly overlapped, and were really, with all
their din and fury, firing at nothing. Further to the left,
and immediately in front of the Light British Division, was a
"gigantic gorge," in which, to quote from a letter of Sir Colin
Campbell, "the enemy had made a large circular redoubt,
protected on each side by artillery on the heights above and
on either side, covered on its flanks and its front by a direct
as well as an enfilading fire. The artillery was supported by
large masses of troops near their guns, and also by other large
masses in rear on the inward slopes of the heights on which
they were posted." This formidable redoubt, armed with 14
heavy guns, was the central point of the engagement. It had
been gallantly taken and as gallantly held; but there had
been bungling somewhere, and the British supports had not
advanced in time. The consequence was that the half-
discomfited Russians, taking heart again, had returned in
renewed strength, and were closely engaged with the far too
slender British force. They poured a literal hail of death
into the British ranks, and the Fusiliers were suffering
terribly—officers being slain in scores, and the men in
hundreds. And still the Russian horde poured on in their
dense columns—including the famous Vladimir column—the
battery behind belched forth its fire, and the holding of the
redoubt had become a literal life and death struggle. The
combat was being waged by twenty men against one, but the
gallant British troops, in this their first contact with the
enemy, were loth to go back. Shot was ploughing through
them, making ugly gaps; shells were shrieking, whistling,
and carrying with them death and destruction; and musket
bullets were finding their billets in many a brave fellow's
heart.

It was a critical time.   Officers were shouting and encouraging their men, and on all sides resounded such cries as—" On lads ; I'll show you the way."   " Close up, close up."   " Stand firm, boys ; now then, steady men, steady." " Forward, forward."   But it was of no use.   The redoubt had to be abandoned, and at length the soldiery were in confusion, here straggling and there huddled together, dispersed along the lower slope, and rapidly falling back. The fate of the battle hung in the balance.   One weak officer had indeed shouted—

" The brigade of Guards will be destroyed ; ought it not to fall back ?"

" It is better, sir, that every man of Her Majesty's Guards should lie dead on the field than that they should turn their backs upon the enemy."

It was Sir Colin Campbell who made the reply, and his words thrilled all by whom they were heard.   For the moment all eyes in the little group of officers, among whom was the Duke of Cambridge, were turned upon him, and everyone read in his calm, stern demeanour the confidence he felt.   This old Brigadier was no novice in the art of war.   He carried the lines of over sixty years on his strongly marked face.   While but a boy of sixteen he had in 1806 been under fire at Vimiera.   There his battalion was formed in open column of companies, and his captain took young Campbell (then an ensign) by the hand, and led him in front of the first company, along which he slowly walked with him for several minutes under fire.   The object was to give the youth confidence, and Campbell, writing of it long years afterwards, said—" It was the greatest kindness that could have been shown me at such a time."   Subsequently he had shared the hardships of the retreat to Corunna, he had led a forlorn hope and been wounded at St Sebastian, and had received

honourable mention in despatches. Besides these services
under the Great Duke, he had been to the West Indies, had
served with distinction in China in 1842, and in India in
1848, achieving a great victory at Chillianwallah. He had
approved himself as brave and capable a leader in his mature

years as he had been full of dash and courage in his youth ;
and although subjected to many disappointments, and often
held back from promotion, when others less fitting but more

influential had been placed over his head, his high qualities as a soldier had at length been recognised, his services in India had been rewarded with a knighthood, and he had now been appointed to the command of a Brigade.

And such a Brigade! His heart swelled with pride as he glanced along the stately ranks of the brave Highland Regiments he was now about to lead into action. Their manly bearing, their picturesque dress, their perfect discipline, their massive, firmly-knit, yet lithe figures, and their apparent eagerness to be led forward, had in them the promise of victory. A Scotsman himself, Sir Colin felt proud of the men under his command. He was, in fact, known to be so proud of them that, says Kinglake, "already, like the Guards, they had a kind of prominence in the army, which was sure to make their bearing in action a broad mark for praise or blame." Sir Colin knew the race from which these men had sprung—knew the glory they had won in the bygone days when their country had rung with their fame. At his side stood the famous Black Watch, eager for the fray, its flag fluttering in the breeze, and bearing the silent but eloquent record—" Egypt," " Corunna," "Fuentes D'Onor," " Pyrenees," " Nivelle," " Nive," " Orthes," " Toulouse," " Peninsula," " Waterloo." Beyond them was the 93rd, which had seen less service than either of its companions; but the glistening, eager eyes of the men conveyed to the old Brigadier that, lead where he might, they would follow. Then on the left—the extreme left of the British line—were the brave Camerons—the 79th—with a record as great, and a history as glorious, as that of the Black Watch itself; and to-day side by side with the premier Regiment they would share the honours and hazards of the conflict.

The Coldstreams, on the Highlanders' right, had advanced, engaged the enemy, then halted, unable to go forward; and one fellow, seeing all going wrong, cried out bitterly—"Let the Scotsmen go on; they'll do the work!"

Now came the turning point of the battle. Sir Colin, with his whole being eloquent of action, turned to the "plumed array" waiting to be led forward, and in stirring tones addressed them :—

"Now, men, you are going into action. Remember this, whoever is wounded—I don't care what his rank is—whoever is wounded, must lie where he falls till the bandsmen come to attend to him. No soldiers must go carrying off wounded men. If any soldier does such a thing, his name will be stuck up in his parish church. Don't be in a hurry about firing. Be steady. Keep silence. Fire low. Now, men, the army will watch us. Make me proud of my High land Brigade!"

Then followed the command, sharp and clear as a trumpet tone, "Forward, 42nd," and as the shrill war notes of the bagpipes rose over the Kourgane Hill the veteran rode— alone—into the strife—through the river and up the slope, where he could see the true nature of the work before him. In a moment his skilled eye had taken in the situation. His three regiments were advancing against twelve—six or seven of them, like his own men, fresh and untouched, and full of the fire and excitement generated by the conflict which had been raging. Standing alone, the General became a mark for the enemy's bullets, and his horse was twice struck; but he quickly measured the enemy's power, and as quickly made his resolve. Already the "superb 42nd" were beside him. Majestically and swiftly they had crossed the river and ascended the height, forming quietly and steadily on the crest, while the 93rd and 79th were also advancing in echelon—a most fortunate disposition, as the result proved. Four battalions, close and compact, faced the Black Watch;

but Sir Colin knew his men, and knew Duncan Cameron, their commander, a man with thirty years of service in the regiment, and to whom the heart of every soldier was

GEN. SIR DUNCAN ALEXANDER CAMERON, K.C.B.

devoted. He came of the best warrior-blood of the Highlands, and was the son of a father who had fought as a veteran in the battles of Colin Campbell's youth.

The 42nd were in line, but as Sir Colin wrote immediately after—"Too much blown to think of charging." A momentary pause to adjust their dressing—during which, to the men's exultation, they saw the two Kazan battalions advancing to meet them, supported by the Vladimir columns, which had retired a good deal battered from the redoubt—then forward went the Highlanders to battle. The order was given to "advance, firing," a manœuvre which Colonel Cameron had derived from his father, Sir John Cameron, and one in which the 42nd had been much practised. It had also the recommendation of being held in high approval by Sir Colin, who had, with the 61st, made it tell with terrible effect on the enemy at Chillianwallah. The soldiers, though young, were cool and steady, and their fire poured close and deadly into the massive Russian ranks. Onward they went, the slender line of picturesquely dressed Scots against the heavy moving battalions opposed to them. The feelings of the enemy were divided between rage and wonder. They had fired volley after volley, but still the bare-kneed line, steady and unwavering, came on. Now and again, after the discharges, the dense smoke rolling between the combatants hid them from one another; then quickly looming out of the cloud appeared again and again the waving plumes nearer and nearer still. But the enemy had the weight of numbers, and moved on to crush the presumptuous invaders. As yet, remember, it was this solitary Highland regiment against the field. And a new danger arose. Heavy enough was the work of the 42nd with the oncoming Kazan columns; but, looking round, Sir Colin saw the left Sousdal battalions on the move, boldly marching to attack his regiment in flank. Instantly perceiving how serious was the situation, he was about to shorten the advancing line of the 42nd by five companies, which he proposed to throw back to meet the new danger; but ere he had given the command his quick eye had seen a better course. The 93rd had reached the crest of the hill, and, "wild and raging," were preparing to dash forward anywhere, at anything, with reckless impetuosity.

"The 93rd in the Crimea," says Kinglake, "was never quite like other regiments, for it chanced that it had received into its ranks a large proportion of those men of eager spirit who had petitioned to be exchanged from regiments left at home to regiments engaged at the war.　The exceeding fire and vehemence and the ever ready energies of the battalion made it an instrument of great weight if only it could be duly held in, but gave it a tendency to be headlong in its desire to hurl itself upon the enemy."

It added to the vehemence of the men's desire to see straight before them an enemy's column, and that column the one threatening the flank of the 42nd.　But Sir Colin once more showed his firm hand.　They would have rushed on headlong as they were, dashing themselves in spray-like particles against the solid rock of the enemy's formation; but he ordered them to halt, then to dress—impressing them in the midst of their fury with the advantages of cool discipline. During this exercise the Brigadier's horse was shot dead under him, for the Russian column had already precipitated matters by opening fire on the 93rd.　The regiment having at length been properly steadied, it got the word to go forward, and then, led by Colonel Ainslie, it, too, dashed into the conflict, and from that moment the flank of the 42nd was safe from any attack by the Sousdal column.　On—in silence, save the discharges of their rifles—the two regiments went straight at the columns before them.　On, yet on, with dauntless courage, closing on the enemy at every step, while death worked havoc on both sides—the loss of the High-landers—slight, indeed, compared with that in the closely-massed columns of the enemy—making no difference to the ardour of the men.　Gaps in the ranks were closed up on the instant, and the lines went swinging forward.　And now, as the din of the fierce fight grew louder, a strange fear filled the hearts of the Russians.　Such men as these—giants in stature and wonderfully apparelled—they had never met

c

before. Their dauntless bearing and their irresistible advance filled the superstitious soldiers of the Czar, first with a

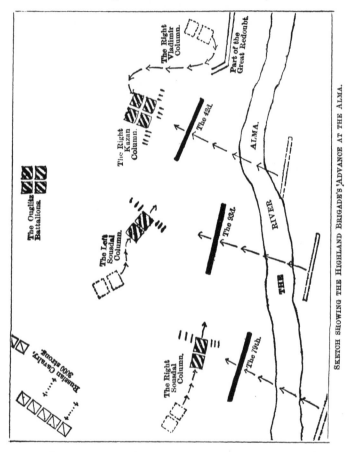

SKETCH SHOWING THE HIGHLAND BRIGADE'S ADVANCE AT THE ALMA.

nameless terror, and then with sheer horror, as they ascribed to the kilted heroes some of the attributes of the denizens of the lower regions. With the stolidness of their nature they for a time stood huddled in their ranks; then they shook, then wavered; then looked around for help—to save them from destruction.

At this juncture they became nerved with fresh courage, for away to their right, and closing on the left flank of the 93rd, appeared the right Sousdal columns—1,500 strong. Could they but close in time, they would double up the slender two-deep ranks of the 93rd like a roll of paper, and then the tide of battle would be turned. Again a critical moment had come, and those watching the struggle from both sides waited, with painful anxiety, the issue of the next few minutes.

Chapter II.

ANOTHER surprise was in store for the distressed Russians. As the right Sousdal regiments advanced exultingly towards the flank of the 93rd, threatening it with destruction, another line of Highlanders appeared suddenly on the top of the hill above the river. "Some witchcraft," writes Mr Kinglake, "the doomed men might fancy, was causing the earth to bear giants. Above the crest or swell of the ground on the left rear of the 93rd yet another array of the tall bending plumes began to rise in a long ceaseless line, stretching far into the east; and presently, in all the grace and beauty that marks a Highland regiment when it springs up the side of a hill, the 79th came bounding forward without a halt, or with only the halt that was needed for dressing the ranks, it advanced upon the flank of the right Sousdal column and caught the mass in its sin—caught it daring to march across the face of a Highland battalion —a battalion already near and swiftly advancing in line. Wrapped in the fire thus poured upon its flank the hapless column could not march—could not live."

It reeled under the hail of bullets, then broke, and began to fall back. From it the flank of the 93rd was now safe, and the spectacle immediately presented to onlookers was the unbroken Highland regiments, now united, extending in grand array for nearly a mile in length, swiftly closing on the disordered Russian columns, which were rapidly giving way. Terror-stricken by this terrible advancing line, and demoralised by the havoc which was every moment being made in their ranks, the Russians finally broke down in sheer despair. From column to column spread the contagion of fear, and a moaning, sorrowful wail burst from their lips. They were undone, and they turned and fled ere the advancing fiends should be at their very throats. The four Ouglitz

GEN. SIR JOHN DOUGLAS, G.C.B.,
*Colonel of the 79th, and Lieutenant-Colonel commanding the Regiment at the
Battle of the Alma.*

battalions standing behind were witnesses of the disaster, and also saw the numerical proportions of the contending parties. As the fleeing masses of their defeated comrades drove towards them, they made a demonstration of advancing to check the Highlanders' onslaught; but in a moment they, too, came within sweep of the murderous fire from the Highland line, united, compact, and moving steadily as ever, and, like the others, they turned and beat a hasty retreat.

The crisis was past; the victory was won; and the Highland Brigade halted—the proud victors who had brought triumph to the allied arms. "As the men," says Kinglake, "looked upon ground where the grey remains of the enemy's broken strength were mournfully rolling away, they could not but see that this, the revoir of the Highlanders, had chanced in a moment of glory." Already the cavalry was after the fugitives, and the guns, which had been dragged up the slope, sent shot after shot among the fleeing and shattered columns. And now the brave old Brigadier rode up to his victorious troops, and, raising his hand, gave the signal to cheer. It was responded to by a thrice-repeated "Hurrah!" which, coming from the throats of lusty, triumphant men, rose above the din of battle, and spread along the slopes of the Alma till the thrill of jubilation was felt from left to right of the allied line. Even the fleeing Russians heard the cry with an amazement which lent wings to their speed.

The battle of the Alma was at an end. Another triumph had been added to the British record of military glory, and to the Highland Brigade and its brave old chief was the triumph due.[*] Most satisfactory of all, the loss of the Highlanders had been comparatively slight—the 42nd, who were longest under fire, lost only 37 in killed and wounded, and the 79th came out of the battle with 2 killed and 7 wounded. The 93rd suffered most severely, having 1 officer (Lieutenant Abercromby), 1 sergeant, and 4 rank and file killed, and 2 sergeants and 40 rank and file wounded. It would be difficult to say which was the more proud—Sir Colin, or his men. Two days after the battle the General wrote from Balaclava an account of the engagement to his friend, General Eyre, in

[*] See Appendix A.

which occurred the following interesting passage :—" It was a fight of the Highland Brigade. Lord Raglan came up afterwards, and sent for me. When I approached him I observed his eyes to fill and his lips and countenance to quiver. He gave me a cordial shake of the hand. The men cheered very much. I told them I was going to ask the Commander-in-Chief a great favour—that he would permit me to have the honour of wearing the Highland bonnet during the rest of the campaign, which pleased them very much ; and so ended my part in the fight of the 20th inst. My men behaved nobly. I never saw troops march to battle with greater *sang froid* and order than those three Highland regiments. Their conduct was much admired by all who witnessed their behaviour. I write on the ground. I have neither stool to sit on nor bed to lie on. I have not had off my clothes since we landed on the 14th." The hardy old soldier did not enjoy the luxury of undressing and going to bed for more than six weeks afterwards.

We need scarcely say that Lord Raglan readily granted his old comrade's request. He felt that Sir Colin had worthily earned a much higher reward than he asked, but he also knew that nothing would so much gratify his heart. It was not—as Sir Colin is said to have remarked on the field—the first victory they had won together, and they felt towards each other that deep friendship which valuable co-operation between men in time of trial and difficulty rarely fails to establish or maintain. The presentation of Sir Colin's bonnet was one of the episodes of the campaign. " The making of the bonnet," writes Mr Keltie, in his History of the Highland Regiments, " was entrusted secretly to Lieutenant and Adjutant Drysdale, of the 42nd. There was a difficulty next morning as to the description of heckle to combine the three regiments of the Brigade. It was at last decided to have one-third of it red to represent the 42nd, and the remaining two-thirds white at the bottom for the 79th and 93rd. Not more than half a dozen knew about the preparation of the bonnet, and these were confined to the 42nd. A brigade parade was ordered on the morning of the 22nd December on the field of Alma, 'as the General was desirous of thanking them for their conduct on the 20th.' The square was formed in readiness for his arrival, and he rode into it with the bonnet on. No order or signal was given for it; but he was greeted with such a succession of cheers, again and again, that both the French and English armies were startled into a perfect state of wonder as to what had taken place."

The brave Highlanders already felt that Sir Colin was one of themselves, and they had given him the love of their warm hearts. Three years afterwards, under a broiling Indian sun, the same regiments marched past him, and Dr W. H. Russell, the celebrated *Times'* war correspondent, who was standing by, wrote in his diary—"They look on him as if he belonged to them, like their bagpipes—a property useful in war." In this sharp and fierce contest with the Russians they had tested each other's value, and had found on both sides that it exceeded the highest expectations. And now that the gallant General had conferred upon his brigade this mark of conspicuous distinction, their enthusiasm for him knew no bounds.

The Russians, when they fled before the fury of the Highland Brigade, crossed the Belbek river, and held to their rear in the direction of Sebastopol, a strongly-fortified town some miles beyond the Alma, at a jutting promontory of the Black Sea coast. Had the French commander agreed to the desire of Lord Raglan to follow them at once the war might have been ended at one fell blow, and many thousands of valuable lives saved. As it was, such a delay occurred in the

advance of the Allies that the Russians were allowed to enter
Sebastopol in safety. By an extremely trying and hazardous,
but well-conceived and brilliantly executed, flank march, the
Allies, however, succeeded in striking across the peninsula to
Balaclava. Then they narrowed their circle around Sebas-

SKETCH MAP SHOWING THE THEATRE OF WAR IN THE CRIMEA.

topol—the object of their expedition, and the prize they had
come to take—and, aided by the sea forces, laid close and
determined siege to the town.

Chapter III.

THE CRIMEA—THE BATTLE OF BALACLAVA.

WITH the engagement on the Alma practically ended the combined action in battle of the Highland Brigade during the Crimean campaign. In the dreary work of the siege, extending over two long and severe winters, the regiments had to endure sufferings of a terrible kind, but their hardships were varied, if not lessened, by the expeditions to Kertch and Yenikale, and Kamara.* When the British army touched the shores of the Crimea, disease was already busy in its ranks. Cold, exposure, wet, want of food, severe fatigue duty, and kindred privations accelerated its ravages. Cholera, dysentery, and other maladies played more havoc than the enemy's bullets. It was a constant, watchful, wearing, harassing toil, without the excitement which strong action supplies. Their share of all this the Highland Brigade, now augmented, bore with unflinching patience. The 71st Highlanders had been sent out. The 72nd also took part in the work in the trenches, and had at that time as adjutant the now distinguished Lieut.-General Sir Archibald Alison, who was commander of the Highland Brigade at Tel-el-Kebir.

But if to the Brigade as a whole there was plenty of risk without a corresponding amount of glory, one of the regiments had another opportunity of obtaining distinction. Sir Colin Campbell, with the rank of Major-General, was appointed Governor of Balaclava, which was the base of the British operations, and the extreme left of the

* See Appendix B.

Allies' position—Colonel Cameron, of the 42nd, having meanwhile succeeded to the command of the Highland Brigade. The 42nd and 79th were drafted to fatigue duty, and to occupy places in the trenches—the 93rd alone of the Highlanders being left in front of Balaclava with Sir Colin.

The position was one of great importance and corresponding defects. Some points of the defence Sir Colin found to be so extremely weak that he regarded them with great anxiety. He set to work, however, with all his energy to repair what was faulty. "The first to rise," says General Shadwell, his biographer, who was at the time a member of his staff, "he was the last to lie down, though engaged on his legs or on horseback every hour of the day in the superintendence of the different working parties, encouraging the diligent, rebuking the indolent, besides visiting at early dawn and nightfall, not only his own posts, but those of the cavalry."

By the 21st of October the defences had been so far completed that Sir Colin expressed his belief that, defective as they were, he could hold them against any but an overwhelming force. The outer line of works was from 2,000 to 2,500 yards in front of the inner line, and was garrisoned by nearly 5,000 Turks, under Rustem Pacha. On the Balaclava plain, and maintaining communication between the outer and inner line of works, was Lord Lucan's command of cavalry, 1,500 strong. The 93rd, strengthened by some other troops, was posted in front of the village of Kadikoi, and nearer Balaclava was a somewhat scattered force of marines. The defences were not completed a day too soon. From the 18th to the 21st the Russian field army lying in front had been seen to make several reconnaissances in force. So strong indeed had they appeared on the latter day that Sir Colin believed an attack on Balaclava imminent, and reinforcements were actually sent to his assistance. The enemy retired, however, for the time, and the reinforcements went back to their posts; but Sir Colin was certain he would not be allowed to remain long at peace. It was an anxious time, and the strain upon the commander with his slender force was very great. He knew the importance of Balaclava to the British—it was their only channel of communication with the outer world, and the way by which they obtained their stores and supplies— and he divined that the Russians would some day make a desperate effort to cut that communication.

He was right. The event occurred sooner even than he expected. On the evening of the 24th the Turkish commander communicated to Sir Colin information from a spy that an attack would be made next day. The night passed quietly, Sir Colin on the alert, as usual, and his men anxious and wary at their posts. Before daybreak the troops were under arms—and not a moment too soon. Breathless and excited riders arrived at headquarters with the intelligence that the enemy, in strong force, were on the move. Immediately all was stir in the British ranks. "Boot and Saddle" blared the trumpets of Lucan's squadron, and immediately the horsemen were riding, "pistols and carbines" loaded, towards the point where they expected to be required.

On Sir Colin, who had been in the saddle since long before daylight, devolved the responsibility of checking the enemy, and as soon as he realised the real nature and magnitude of the attack, he decided on his course of action. He immediately rode up to, and placed himself by the side of, the gallant 93rd, who impatiently awaited his orders. The scene presented to him in the dim morning was of the most exciting kind, and one calculated to try even the strongest nerves. Debouching from the hills in front were great bodies of the enemy. They were of all arms—infantry, cavalry, and artillery—and were moving rapidly in the direction of the

Turkish redoubts. As compared with the slender forces holding Balaclava, the numbers were indeed overwhelming— 25 battalions of infantry, 34 squadrons of cavalry, and 78 guns—the whole amounting to about 24,000 men. Lucan's cavalry had gone to assist the Turks in the redoubts, against whom it was now clear the attack was primarily directed; and Sir Colin and the 93rd stood calmly at their post to watch the course of events. On came the Russians with irresistible force—covering their attack with 30 guns—and soon it was seen that the Turks in No. 1 Redoubt, subjected to a terrible fire, were giving way. In a few minutes more the redoubt was carried, and its defenders were in flight. With this in possession of the enemy, it was impossible for the others to hold out, and in a short time they were carried in detail, and the Turks were seen fleeing from their posts, and making, in a straggling mass, for Balaclava, between which and themselves the 93rd was drawn up. Seeing the steady array of the Highlanders, the fugitives ran towards the line, and were hastily formed on either flank.

And now the victorious Russians, being in full possession of the redoubts, advanced in force into the gentle valley which lay between themselves and the Highlanders, who occupied a piece of slightly rising ground. In their thousands they moved forward, and their artillery coming within range, opened fire so successfully that one or two of the Highlanders and some of the Turks were wounded. Seeing this, Sir Colin retired his men behind the crest of the hill, and as they lay down he watched the development of the Russian movement. It was quickly revealed to him, for, as he watched, four squadrons of the enemy's cavalry, suddenly detaching themselves from the main body and heading straight for the 93rd, galloped forward at the charge.

A critical moment was at hand, and one in which the

chances were entirely in favour of the advancing horsemen. The force in Campbell's hand was slender indeed when the task before it is considered. Formed in line, only two deep, were 550 of the 93rd, and about 100 invalids whom Colonel Daveney had drawn up on the Highlanders' left. In addition were the Turks already mentioned, on whom, however, no reliance could be placed. But the General had confidence in his Highlanders, and to show it he rode down the line and said—

"Now, men, remember there is no retreat from here. You must die where you stand."

The response was decided and cheerful—

"Ay, ay, Sir Colin; an' need be, we'll do that."

It was John Scott, the right-hand man of No. 6 Company, who spoke, and others took up and shouted forth the reply.

Sir Colin immediately ordered the Highlanders forward to the crest of the hill, and the men obeyed with an impetuosity which suggested a desire to rush on and charge the advancing enemy. But this would have ruined all, and as they sprang forward Sir Colin, with his temper at fever heat, was heard fiercely shouting—

"Ninety-third! Ninety-third! D——n all that eagerness!"

"The angry voice of the old man," says Kinglake, "quickly steadied the line." And now came an exhibition of quiet resolute courage such as soldiers have seldom displayed on the field of battle. Discarding the usual method adopted by infantry of receiving cavalry in square—not even troubling himself to throw his men into fours—Sir Colin awaited the onslaught with his "thin red line" of two deep. As the thunder of the furiously galloping horse and the cries of the riders fell upon the ears of the Turks, huddled on the flanks of the 93rd, they quickly broke, and once more ran to the rear in utter affright, holding out their hands to the ships in the roadstead, and crying out, "Ship, ship." But the 93rd stood firm as the unshaken rock. Nearer and nearer came the cavalry, their swords, lanceheads, and bright helmets glittering in the now clear morning light. Their pace was furious—General Wolseley calculates it at three hundred and fifty yards a minute—the ground seeming literally to fly beneath their feet, and the manner in which they brandished their weapons showed the fierceness of their desire for the combat. But combat was hardly to be expected, for "that thin red streak, tipped with steel," might have been regarded as no greater an obstacle than a fence of furze. And coming on behind the leaders were squadron after squadron, like, says James Grant, "the successive waves of a human sea."

It was a terrible trial for men to stand unmoved and watch this raging avalanche hurling itself against them. "In other parts of the field," says Dr Russell, of the *Times*, who saw the action, "with breathless suspense every one waited the bursting of the wave upon the line of Gaelic rock." But the time for action had come. Suddenly a word of command rang out sharp and clear, and the rifles of the 93rd were levelled at the advancing foe. The plumed heads drooped as the regulation three seconds were spent in taking careful aim. Then flashed out from flank to flank a withering volley, which sent dismay into the enemy's ranks, caused them to reel, stagger, stumble, and recoil. Their headlong course was checked, and as they tried to extricate themselves from the wild confusion into which they had been thrown, the cool Highlanders, calmly as if on parade, brought their butts to the ground and reloaded. A detachment headed off from the main body of the enemy, and moving to the left, attempted to outflank the 93rd. "That man knows his business, Shadwell," said Sir Colin to a staff officer beside

him; and also knowing his business, Sir Colin wheeled a
portion of his men to the right to meet the emergency. The
movement was successful. One more volley, and the discomfited horsemen were galloping back in full retreat.
"Well done, brave Highlanders," shouted the spectators,
as they for a moment breathed again. A great end had been
achieved, a marvellous feat in warfare accomplished. The
weakest point in the defence of Balaclava had been maintained,
and the Russian opportunity was lost. General Burroughs,
who was at the time Lieutenant of No. 6 Company, states in
"The Records of the 93rd Regiment," that a party of British
officers were afterwards informed by Russian officers who were
in the engagement that "few of us were killed, but nearly every
man and horse was wounded." Many horses were killed.

"The advance of the Russian squadrons," writes Mr
Kinglake, "marked what might well seem at the moment to
be an ugly if not desperate crisis in the defence of the
English seaport. Few or none at the time could have had
safe grounds for believing that, before the arrival of succours,
Liprandi (the Russian commander) would be at all once
stayed in his career of victory, and in the judgment of those,
if any there were, who suffered themselves to grow thoughtful, the whole power of our people in the plain and the port
of Balaclava must have seemed to be in jeopardy; for not
only had the enemy overmastered the outer line of defence,
and triumphantly broken in through it, but also having a
weight of numbers, which for the moment stood as that of
an army to a regiment, he already had made bold to be
driving his cavalry at the very heart of the English resources.
If, in such a condition of things, some few hundreds of
infantry men stood shoulder to shoulder in line confronting
the victor upon open ground, and maintaining from first to last
their composure, their cheerfulness, nay, even their soldierly

mirth—they proved themselves brave men by a test which was other than that of sharp combat, but hardly less trying. And the Highlanders," the historian continues, "whilst in this joyous mood, were not without a subject of merriment; for they saw how the Turks in their flight met a new and terrible foe. There came out from the camp of the Highland regiment a stalwart and angry Scotch wife, with an uplifted stick in her hand; and then, if ever in history, the fortunes of Islam waned low beneath the manifest ascendant of the Cross; for the blows of this Christian woman fell thick on the backs of the Faithful.* She believed, it seems, that, besides being guilty of running away, the Turks meant to pillage her camp, and the blows she delivered were not mere expressions of scorn, but actual and fierce punishment. In one instance she laid hold of a strong-looking, burly Turk and held him fast until she had beaten him for some time, and seemingly with great fury. She also applied much invective. Notwithstanding graver claims upon their attention, the men of the 93rd were able to witness this incident. It mightily pleased and amused them."

A few days later Sir Colin and his men had the gratification of receiving the thanks of the Commander-in-Chief for their heroic deed. It was in the form of a general order, and was in the following terms:—"The Commander of the Forces feels deeply indebted to Major-General Sir Colin Campbell for his able and persevering exertions in the action in front of Balaclava on the 25th inst., and he has much pleasure in publishing to the army the brilliant manner in which the 93rd Highlanders, under his able directions, repulsed the enemy's cavalry. The Major-General (Sir Colin) had such confidence in this distinguished regiment that he was satisfied that it should receive the charge in line, and the result proved that his confidence was not misplaced."

D   * See Appendix C.

The phases of the battle which followed the repulse of the Russian cavalry by the 93rd do not come within the scope of our task.   Yet we may mention that deeds of heroism, furnishing an imperishable page of British history, were immediately after performed by the Light and Heavy Brigades, who made the famous charges through the Valley of Death. Many, very many, lives were lost; but the survivors earned undying glory, and the fallen were stricken down in a combat which, on their side, for fierceness and hopelessness has no parallel in modern history.

Chapter IV.

WITH the engagement at Balaclava the duties of the Highland Brigade in the Crimea did not end, but the men had no further chance of displaying conspicuous heroism in battle. A great opportunity was lost on the 9th of September 1855. On the previous day the Brigade had been detailed as supports in the grand assault on the Redan, but had not engaged in the fight. Next day the assault was to have been renewed, with the Highland Brigade as the storming party; but the enemy evacuated the position, and the hazardous service—in which the men were eager to engage—was not required.

An incident connected with the evacuation of Sebastopol is recorded in Fullarton's "Highland Regiments" by Mr Keltie, and its importance makes it worth repeating :—
"About midnight on the 8th, the Russian fire having previously ceased, and everything appearing unusually quiet, Lieutenant W. M'Bean, the Adjutant of the 93rd, left the advanced trench, and, approaching the Redan, was struck with the idea that it was deserted by the Russians. He accordingly gallantly volunteered to enter it, which he did with a party of ten volunteers of the Light Company, under Lieutenant Fenwick, and a like number of the 72nd under Captain Rice. They found no one in the Redan but the dead and wounded left after the assault. The party, however, had a narrow escape, as an explosion took place in the Redan shortly after."

On the first anniversary of the battle of the Alma, Sir

Colin Campbell had an opportunity of expressing his admiration for the Highland Brigade. On that day the first presentation of Crimean medals was made, and Sir Colin issued a stirring address to the troops, in which he said :—

"HIGHLAND BRIGADE,

"On the first anniversary of the glorious battle of the Alma our gracious Sovereign has commanded the Crimean medal to be presented to her gallant soldiers who were the first to meet the Russians and defeat them on their own territory. The fatigues and hardships of that year are well known, and have greatly thinned our ranks since the day we scaled the Alma heights together. To that day Scotchmen can look with pride. For your deeds upon that day you received the marked encomiums of Lord Raglan, the thanks of the Queen, and the admiration of all. Scotchmen are proud of you. I, too, am a Scotchman, and proud of the honour of commanding so distinguished a Brigade ; and still prouder that through all the trying severities of the winter, its incessant labours and decimating disease, you have still maintained the same unflinching courage and energy with which your discipline, obedience, and steadiness, in whatever circumstances you are placed, make you so unrivalled (and none more so than the oldest regiment of the Brigade), and your Commander confident of success, however numerous and determined your foe.

"I cannot conclude without bringing to your minds that the eyes of your countrymen are upon you. I know you think of it, and will endeavour by every effort to maintain your famed and admirable discipline ; also that your conduct in private equals your prowess in the field ; and when the day arrives that you are no longer required in the field, welcome arms will be ready to meet you with pride, and give you the blessings your deeds have so materially aided to bring to your country. And in after years, when recalling the scenes of the Crimea by your ingleside, your greatest pride will be that you, too, were there, and proved yourself a worthy son of sires who, in bygone days, on many a field added lustre to their country's fame. . . .

"Remember never to lose sight of the circumstance that you are natives of Scotland; that your country admires you for your bravery; that it still expects much from you; in short, let every one consider himself an hero of Scotland. It is my pride, and shall also be my boast amongst the few friends which Providence has left me, and those which I have acquired, that this decoration of the Order of the Bath which I now wear has been conferred upon me on account of the distinguished gallantry you have displayed."

The old man's love for his Highlanders is revealed in every line of the above. It is painful to think that so gallant a soldier and so noble a man should have so often been pushed aside in the struggle for place. In the Crimea he had reason—and perhaps good reason—to think that his services were not regarded in certain influential quarters with the favour they deserved; and he found himself more than once left out in the cold, when his experience and his courage entitled him to a leading voice in deliberation, and a foremost place in action. This treatment was so painful to him that he determined to retire from the army; and obtaining leave of absence, quitted the Crimea for England early in November 1855. After firmly demanding and obtaining explanations from Ministers, and assurance of the consideration to which his services and talents entitled him, his love for soldiering prevailed over his pride, and he returned two months later to his post at the seat of war.

Before leaving the subject of the Crimea, one or two incidents illustrating the vicissitudes of the men in the trenches may be given. As a rule, it was exhausting, miserable work. Yet the British soldier is very slow to grumble at the round of duty, whatever the hardships of that duty may be ; and during the siege of Sebastopol, which proved so disastrous to the health and lives of the besiegers, the troops bore up with a patient heroism that won for them the respect even of their enemies. For the most part, life in the trenches was an arduous thing ; but it was not without its brighter side. If risks had to be run and accidents deplored, there were also dangers to be escaped and congratulations bestowed. If the messenger of death was often in their midst, the guardian angel was also present ever and anon snatching one and another brave fellow from the line of the bullet or the radius of the bursting shell. There was an irrepressible spirit of good humour pervading the ranks, conquering the disposition to be melancholy over the sufferings and death of their comrades and the constant danger to themselves. There was likewise the steadfast hope of getting once more, as at the Alma, Balaclava, and Inkerman, to close quarters with the enemy, and giving them another proof of true British mettle.

As an example of the stoical fortitude displayed by our soldiers, Mr Keltie gives what he truly terms "a striking and intensely pathetic reminiscence of the campaign." It was furnished him by Lieutenant-Colonel Clephane, and is of so affecting a nature that we quote the statement entire. Our only regret is that we cannot furnish our readers with the brave fellow's name :—

" Shortly after the opening of the bombardment of Sebastopol, the 79th Highlanders furnished a party for trench duty, consisting of about 150 men under the command of a field officer, and accompanied by a similar number detailed from the brigade of Guards. They marched for the post of duty shortly after daybreak, taking the well-known route through the 'Valley of Death,' as it was called. At that time a foe, more dreaded than the Russians, had persistently dogged the footsteps of the army, never attacking it in force, but picking out a victim here and there with such unerring certainty, that to be sensible of his approach was to feel doomed. The glimmering light was insufficient for making out more than the dark body of men that moved silently along the above gloomy locality in columns of march four deep; but as the sun approached nearer the horizon, and the eye became accustomed to the glimmer, it was seen that one man was suffering under pain of no ordinary nature, and was far from fit to go on duty that morning. Indeed, on being closely inspected, it became evident that the destroyer had set his seal on the unfortunate fellow's brow, and how he had mustered determination to equip himself and march out with the rest was almost inconceivable. Upon being questioned, however, he persisted that there was not much the matter, though he owned to spasms in his inside and cramps in his legs, and he steadily refused to return to camp without positive orders to that effect, maintaining that he would be better as soon as he could get time to 'lie down a bit.' All this time the colour of the poor fellow's face was positively and intensely blue, and the damps of death were standing unmistakably on his forehead. He staggered as he walked, groaning with clenched teeth, but keeping step and shifting his rifle with the rest, in obedience to every word of command. He ought probably to have been at once despatched to the rear, but the party was now close to the scene of action (Gordon's Battery), the firing would immediately commence, and somehow he was for the moment forgotten. The men took their places lining the breastwork, the troops whom they relieved marched off, and the firing began and

was kept up with great fury on both sides. All at once a
figure staggered out from the hollow beneath the earthen
rampart where the men were lying, and fell groaning upon
the earth a few paces to the rear. It was the unfortunate
man whose case we have just noticed. He was now in the
last extremity, and there was not the ghost of a chance for
him in this world; but three or four of his comrades
instantly left their place of comparative safety, and surrounded
him with a view of doing what they could to alleviate his
sufferings. It was not much; they raised him up and rubbed
his legs, which were knotted with cramps, and brandy from
an officer's flask was administered without stint. All in vain,
of course; but curiously enough, even then the dying man
did not lose heart or show any weakness under sufferings
which must have been frightful. He was grateful to the
men who were busy rubbing his agonised limbs, and expressed
satisfaction with their efforts, after a fashion that had even
some show of piteous humour about it. ' Aye,' groaned he,
as they came upon a knot of sinews as large as a pigeon's
egg, ' that's the *vaygabone*.' It became evident now that the
best thing that could be done would be to get home to camp,
so that he might at least die beyond the reach of shot and
shell. The open ground to the rear of the battery was swept
by a perfect storm of these missiles; but volunteers at once
came forward and placed upon one of the blood-stained
litters the dying man, who, now nearly insensible, was
carried back to his tent. This was effected without casualty
to the bearers, who forthwith returned to their post, leaving
their unfortunate comrade on the point of breathing his last."

Examples such as the above attest the grit of the men who
laid siege to the Russian stronghold, and so many of whom
yielded up their lives in the struggle. The task had also its
lighter side, and this is partially exhibited in the following

notes of experiences of life in the trenches, kindly furnished
by an old 72nd Highlander, who passed through the ordeal,
and who is at the present moment Sergt.-Major of one of the
Dundee Rifle Corps :—" I remember," he says, " a glorious
Sunday afternoon in the camp before Sebastopol. Along
with a number of others, I had been ordered to parade for
duty in the trenches, and after the usual preliminaries, we
were marched to relieve the men who had been on duty
during the previous twenty-four hours. A halt was made in
the valley, known after the glorious charges of the Light and
Heavy Brigades as the ' Valley of Death,' and the men were
permitted to fall out for a time till the arrangements for
entering on trench duty had been completed. During the
halt the men grouped together, and some of us got to chatting
about home, and the old folks there. We speculated as to
what they would be doing, and what they would be thinking
of us. An old schoolfellow and myself by-and-by got
together, and looking on the scene around us, we were both
forcibly struck with its resemblance to a nook at the bottom
of the Lomond Hills. This kept our minds for the time
rather firmly fixed on home, and as both my comrade and
myself had abandoned fairly good prospects in life to join
the army and meet the hazardous chances now before us, our
thoughts took a sad rather than a cheerful turn. This was a
frequent experience during the dreary, monotonous months
of siege duty. But immediately we got within range of the
enemy's fire all such thoughts were driven out of our heads.
An indescribable feeling often took possession of me, and I
believe all young soldiers are subject to it—that is, when
aware they are in danger the desire seizes them to rush
forward nearer to its source, as if to grapple with it. There
was a good deal of excitement in getting both in and out of
the trenches, and we had to be careful how we exposed our-

selves to the enemy. We had to dodge their shot and shell, and, to vary the excitement a little, some of our own gunners in the batteries behind us, while firing on the enemy over our heads to cover our movements, would sometimes lose their range and send a round shot into a group, or cause a shell, intended for the enemy, to burst over our heads. It was, indeed, a very stirring two hours before we got fairly settled in our places for the night.

"On the night to which I am alluding we had rather a startling experience. We happened to be close to a place called the Quarries, and in the Quarries was a working party of the 42nd. I was outside with my own corps; but an old comrade of mine, who had volunteered and joined the 42nd at the commencement of the war, was engaged with the working party. Hearing him talk, I called him by name, and, after a few minutes, he came out to me. We were chatting together when we observed a 13-inch shell, that had been fired from a mortar about Sebastopol, attain its elevation almost over where we were standing. 'It will fall here,' I remarked. 'No,' said D——; 'it is meant for the working party inside; let us stand still, and we'll be all right.' Down came the shell, the fuse hissing fearfully, and struck on the top of the embankment where the working party were busy, rested a moment, then rolled fizzing down the embankment to our very feet. D—— ran round the corner of the embankment, which was but a few steps off, calling on me to follow. But I was petrified, and could not move, so out he came, caught me by the sash, and was dragging me in when we both fell and rolled over together. At the same moment the shell burst, and in an instant we were covered with dirt and sand. But not a word was spoken—for an age, as it seemed. At length said D——, 'Are you hurt?' 'I don't know,' I answered solemnly. 'Get up at once, then,' was his

unceremonious reply. 'If you don't know whether you
are hurt or not, there's not much the matter with you.' He
was quite correct. We had just had a most providential
escape. But all had not been so fortunate. A nice young
lad, a piper in the regiment, who had been only a few yards
from us, was literally blown to atoms, and two other poor
fellows were wounded. D——, it is worthy of mention, has
since had his name enrolled on the record of true British
heroes, having received the Victoria Cross for valour during
the Indian Mutiny.

"Some very ludicrous incidents also took place. When a
man paraded for duty in the trenches, he took his rations for
twenty-four hours along with him. These consisted of salt
beef or pork, biscuits, coffee, sugar, &c. In the trenches
they had to be cooked in the best way we could manage.
On arriving at our destination, we were formed in single
line, with slight intervals between the men, and then ordered
to number by threes, when all the Nos. 1 were considered on
duty for two hours, then the Nos. 2, and then the Nos. 3.
This gave the off numbers an opportunity of getting a sleep,
for, strange as it may appear, we used to sleep soundly, even
in the trenches, with the shot and shell whizzing and hissing
about our ears. To cook our food we required a fire, and to
make a fire we required fuel, and the latter we had to look
about for. One favourite way of securing fuel was to obtain
the wooden case of a Moorsom shell, which was sufficient to
cook all the food one was likely to have. When a fellow got
hold of one of these he considered himself fortunate, and
generally took special care that he did not lose or get it
stolen. One night one of my squad had got hold of a case,
and had laid it on the gabion beside him. Much as the
poor fellow prized his treasure, and anxious as he was to
keep it secure, he had succumbed to the demands of nature

and fallen sound asleep. He was lying thus at the bottom of
the gabion when a round shot from Sebastopol struck the
outside of the trench, and down fell the empty case on
Sandy's head. Startled and stunned he roared out that he
was done for, and a number of his comrades gathering round,
he was anxiously asked where he had been hurt. His face
was bespattered with blood, and he was blinded with dirt;
but though he felt he had been seriously wounded, he could
not tell where the worst part was. A stretcher was procured,
and Sandy was laid upon it. He was then carefully carried
back half a mile to where the doctors had improvised a field
hospital. He was taken in, laid on the rude table, and his
face carefully sponged. The doctor, glancing at the wounded
man, appeared to clearly understand the case. ' Are you in
much pain, my man ?' he asked. Sandy only groaned out,
' Oh !' ' Come, come,' says the doctor, ' keep quiet ; try and
sit up.' ' Oh !' again groaned Sandy. The doctor then
called on the orderly to bring in a little brandy, which he
gave to Sandy, and asked him to try and drink it, a request
with which Sandy most readily complied. Seeing this, the
doctor cried out, ' Oh !' in imitation of Sandy, adding, ' Get
down off the table ; there is nothing the matter with you !'
Poor Sandy did look crestfallen to find that he had not been
wounded. He was loth to believe that his face had only been
scratched by the precious case falling on his head. On his
return to camp he was accompanied by the men who had
carried him to the hospital, and their good-natured chaff at
his expense was rather trying to his temper—one of them
asserting that it was rather a cheeky joke to have them carry
him half a mile to get a glass of brandy. As soon as Sandy
reached his post the unlucky box caught his eye, and seizing
it, he flung it from him, declaring he would rather starve
than have a fire of it.

"On another occasion a jolly little fellow, nicknamed 'Fuchsia' Fraser, was boiling his coffee in his mess tin on his little fire. When he tried to remove the tin he found it too hot for him to lift. He stepped to where his rifle was lying, and was in the act of drawing his ramrod to put through the handle of the tin, when a round shot fell smash right into the fire, playing havoc with both it and the mess tin. For a moment 'Fuchsia' stood speechless, regarding his ruined mess, then burst out—'Confound the thing; if I had only been a little sooner.' At the moment he entirely forgot that he had just saved his life by being away from the fire."

The Sergt.-Major also furnishes a narrative of a narrow escape made by himself and a squad of the 72nd, who lost their way in the trenches, and almost landed within the enemy's lines. He describes the cannonade on the morning of the 18th June—the day of the grand assault on the Malakoff and Redan—as most terrific; it seemed as if the iron hail was coming from heaven itself. The day, he says, he has always considered the longest of his life—one that he thought would never come to an end. Before mid-day he had consumed all his rations, and it was almost midnight before the men got back to their tents. He was, however, fortunate in getting in amongst a lot administered to by 'Jocky Garden,' who gave them stewed pork and biscuits. Jocky was a favourite with the men ; if not on trench duty, he was always thinking of and preparing for his comrades who were on duty; and when it was Jocky's turn to go on duty his comrades, says the Sergt.-Major, did not forget him.

In 1857, a handsome monument was erected in the Dean Cemetery, Edinburgh, "To the memory of Colonel the Honourable Lauderdale Maule, Lieut.-Colonel E. J. Elliot, Lieut.-Colonel James Ferguson, Captain Adam Maitland, Lieut. F. A. Grant Lieut. F. J. Harrison, and Dr R. J.

Mackenzie ; also, 369 non-commissioned officers and men of the 79th Highlanders who died in Bulgaria and the Crimea, and who fell in action during the campaign of 1854-55."

MONUMENT IN DEAN CEMETERY, EDINBURGH.

This long death-roll illustrates the havoc played by disease in the British ranks.

At the conclusion of our last Chapter we mentioned that Sir Colin Campbell returned to the Crimea after his visit to England. He had been so kindly received, and had been so mollified by the high esteem in which he found himself held in influential circles, that he returned to the Crimea prepared to place himself "under a corporal" if Her Majesty's service required it. But he reached the seat of war too late. Sebastopol had already fallen, and peace had been secured.

As Sir Colin then viewed it, war with him was at an end, and he would henceforth follow the paths of peace. He determined to return home. But the love for his Highland Brigade burned more strongly than ever in his heart. Ere he departed he assembled the 42nd, the 79th, and the 93rd together, and took leave of them in the following address, which is perhaps the most pathetic and affecting ever offered by a commander to his men :—

"Soldiers of the 42nd, 79th, and 93rd! old Highland Brigade, with whom I passed the early and perilous part of this war, I have now to take leave of you. In a few hours I shall be on board ship, never to see you again as a body. A long farewell! I am now old, and shall not be called to serve any more, and nothing will remain to me but the memory of my campaigns, and of the enduring, hardy, generous soldiers with whom I have been associated, whose name and glory will long be kept alive in the hearts of our countrymen. When you go home, as you gradually fulfil your term of service, each to his family and his cottage, you will tell the story of your immortal advance in that victorious echelon up the heights of Alma, and of the old Brigadier who led you and loved you so well. Your children and your children's children will repeat the tale to other generations when only a few lines of history will remain to record all the enthusiasm and discipline which have borne you so stoutly to the end of this war. Our native land will never forget the name of the Highland Brigade, and in some future war that nation will call for another one to equal this, which it can never surpass. Though I shall be gone, the thought of you will go with me wherever I may be, and cheer my old age with a glorious recollection of dangers confronted and hardships endured. A pipe will never sound near me without carrying me back to the bright days when I was at your head, and wore the bonnet which you gained for me, and the honourable decorations on my breast, many of which I owe to your conduct. Brave soldiers, kind comrades, farewell!"

The men were deeply moved by the eloquence of their brave old leader. Like him, they regarded the separation as final, and they freely showed their affection and regret.

But the end of their glory had not come yet. The future was pregnant with a great trial for Britain. In another and more distant land her strong arm was called into action, and a mighty struggle had to be waged to preserve her power. In that struggle Sir Colin and the Highland Brigade were called upon to take an arduous, and, as it proved, an honourable part.

Chapter V.

THE Persian Campaign of 1856-57 was comparatively of so little importance that we might have dismissed it in a paragraph, but from the circumstance that engaged in it were a regiment of Highlanders and two British Generals—Outram and Havelock—who in the Indian Mutiny gained a glory which shall never fade. Persian aggressions in the direction of Afghanistan caused the Governor-General of India, on November 1, 1856, to declare war against the Shah. Troops under General Stalker were immediately despatched from Bombay to Bushire, where a landing was effected. That town quickly surrendered, and the British force pitched its camp there, to wait until the arrival of reinforcements should permit an advance in strength to strike a blow that would be felt.

Among the reinforcements was the 78th Highlanders—or "Ross-shire Buffs"—a regiment that had already gained great distinction in warfare, and one which was known to be composed of a large proportion of genuine Highlanders—men with good bone, muscle, and sinew, and with brave, undaunted hearts. The regiment had seen many vicissitudes, and had always acquitted itself with honour. Its flag bore the inscriptions "Assaye," "Maida," "Java," and its record told how gallantly in many other engagements, and under many

B

other climates, it had sustained the brunt of battle. It was over 800 strong, and commanded by Colonel Stisted, a brave and capable officer, who was ably seconded by Major M'Intyre.

To the command of the division which comprised the 78th was subsequently appointed an officer who, like Sir Colin Campbell, had great confidence in the Highlanders, and one in whom all who knew him or had served with him could place unbounded reliance. In all the long list of Britain's warrior heroes there is no more honoured name than that of Sir Henry Havelock, and no braver man ever drew the sword for Sovereign and country. Originally intended for the Law, he had received a liberal education, and was a cultured man, whose strong piety governed all his actions. He was now over sixty, and his military experience of more than thirty years had been gained in India during many hazardous and trying wars. Besides possessing all a true soldier's aspirations for honour and distinction, his circumstances were such that the emoluments of a responsible command had a strong attraction, and he had eagerly accepted the post when it was offered. He quickly recognised the high qualities of the 78th, and when the opportunity came placed them at the point of danger, which is always to a soldier the point of honour.

In supreme command of the expedition was another man of outstanding qualities—one of the most distinguished military administrators of his time—General Outram. He, too, had gained his laurels in India—and a singular incident in connection with the appointments of himself and Havelock may be mentioned. Havelock, who had been consulted by the military authorities in India at the outbreak of the war, as to the fittest man to act as Commander-in-Chief, instantly named Outram. Outram, on the other hand, who was in England at the time, and knew

nothing of this, instantly on receiving his appointment as
Commander-in-Chief, advised that General Havelock should

From engraving in " *Goldsmid's Life of Outram.*"

receive the command of the second division. Outram, although
of rather insignificant presence, was possessed of indomit-

able courage and an iron will. In addition, his administrative and diplomatic abilities were of the very highest order—and so utterly unselfish was he that he had already earned the title of "the Bayard of India"—the single line which, at the suggestion of Dean Stanley, serves as the inscription on the slab which marks his grave in Westminster Abbey. When but a boy Outram had displayed many of the qualities which marked him as a man. He had stood on the seashore, when his hand had been caught by a crab, and the blood was streaming from it, and, holding the creature aloft, watched with set teeth, but without cry or grimace, until its own weight compelled the creature to relax its hold and fall to the ground. "I thought he'd get tired at last," he coolly remarked, as he quietly wiped his bleeding fingers. At Aberdeen, where he was educated, he became a favourite with the soldiers, and an equal favourite with the sailors. One day, however, the sailors having mutinied, the soldiers were sent against them with muskets loaded, and bloodshed was expected. All Aberdeen was excited, and one of the incidents long remembered was that when the soldiers came close to the sailors, who were drawn up in a body on the quay, little Jamie Outram—only thirteen—was seen pacing about "like a tiger in his den," between the expected combatants, "protecting his sailor friends from the threatening muskets, and resolved to receive the fire first, if firing there was to be." Matters, however, ended peacefully, and Outram was not called on to make any further display of his boyish heroism. There had been some thought of training him for the Church; but Outram revolted. "You see that window," he said to his sister; "rather than be a parson, I'm out of it, and I'll 'list for a common soldier." And into the army he had gone at seventeen, as a cadet in the Indian service. Then had come the opportunity of proving that "he had the heart and courage of a giant in the body of a pigmy." Yet, diminutive-looking as he was, he was no weakling. His constitution was of iron. He had determined to conquer climate and disease, and had been successful. Cholera, fever, and dysentery had attacked him again and again, and he had emerged from the illness to labours and duties more fatiguing and arduous than before. His military and diplomatic services had been of the most brilliant kind. He had pursued Dost Mahomed through the almost inaccessible wilds of Afghanistan with a daring and rapidity which had won for him the admiration alike of India and of Britain. In the campaign which terminated with the annexation of Scinde he had distinguished himself by conspicuous valour, and had divided the amount of his prize-money among the benevolent institutions of India. Outram had the further advantage of being quickly able, by his chivalrous bearing and generous appreciation of honest service, to popularise himself with all ranks under him.

Thus ably commanded, the expedition moved forward to success. The first operation was a march against the enemy's position, forty-six miles from Bushire, which, upon being reached, was found partially abandoned. After destroying everything that could be found, Outram ordered a return to Bushire. The march back was commenced early in the evening of February 7th. For a few hours everything went well; but at midnight the troops were startled by the rattle of musketry, and the booming of heavy guns in the rear. The enemy had formed an ambuscade on the line of march, and attacked the British column. Soon the whole force was engaged in a fierce skirmishing battle. Captain Hunt, of the 78th, the historian of the campaign, thus describes the affair :—" Horsemen galloped round on all sides, yelling and screaming like fiends, and with trumpets and bugles making as much noise as possible. One of their buglers had the audacity to go close to a skirmishing body of the 78th, and sound first the 'cease fire,' and afterwards 'incline to the left,' escaping in the dark. Several English officers having, but a few years since, being employed in organising the Persian troops, accounted for the knowledge of our bugle calls, now artfully used to create confusion. The steadiness of the men was most admirable. . . . . The horsemen of the enemy were at first very bold, dashing close up to the line, and on one occasion especially, close to the front of the 78th Highlanders; but, finding that they could occasion no disorder, this desultory system of attack gradually ceased, and the arrangement of the troops for the night was effected under nothing more serious than a distant skirmishing fire."

During the night the enemy opened a heavy fire with the big guns on the camp, but fortunately did little harm. In the morning their punishment came. They were found strongly posted at the village of Kooshab. The British force at once moved against them, the 78th being in the leading line. In front of the enemy's centre were two mounds, which were used as redoubts, behind which cannon were placed. These had to be taken, and the Highlanders advanced in splendid style—at a rapid pace getting over the rising ground covered by the enemy's guns, and closing with the astonished Persians before they could get out of the way. The slaughter of the enemy was terrific, more than 700 dead being left on the field, while the loss of the British was comparatively trifling. For this affair the word "Kooshab" was ordered to be inscribed on the colours of the 78th, and Colonel Stisted, their commander, was mentioned with honour in Outram's despatches.

The march was afterwards resumed. It was a hazardous and painful one—through swamps and mire, and under a

and painful one—through swamps and mire, and under a

pelting rain—yet at the end of fifty hours from its commencement, the troops had marched forty-four miles, and fought
and defeated the enemy twice—a very creditable accomplishment indeed.

The concluding general engagement of this brief campaign
took place in March. The enemy had strongly entrenched
themselves at the village of Mohumrah, on the banks of the
river Euphrates, and an expedition was despatched to drive
them out. Among the troops detailed was the 78th, which
mustered its full strength for the service. For months the
Persians had been strengthening the position. They had
erected strong batteries and earthworks, mounted with heavy
ordnance, which were so skilfully placed that the whole course
of the river was swept by their fire. Nothing had been left
undone to render the position safe from attack, and the
enemy deemed it impossible that any craft could pass up the
river through their fire. They mustered 13,000 men of all
arms, had 30 guns, and were commanded by Khander Meerza
—the Shah-Zada, a prince of the blood, and a man of some
military repute. The plan of the British operations was
this—To attack the enemy's batteries with steamers and ships
of war, and when the fire was nearly silenced, to pass up the
river rapidly with the troops in small steamers and towing
boats, land the force above the northern forts, and immediately
advance upon the entrenched camp.

The British expedition, consisting of 4886 men, of which
nearly two-thirds were Indian troops, sailed up the river in
fine order, Havelock commanding and accompanying the
leading division. On arriving within three miles of the
enemy's southern battery the vessels were anchored, and at
daybreak on the 26th March a heavy fire was opened on the
position from mortar rafts and the war vessels. By nine o'clock
the batteries were almost silenced. An incessant shower of

shot and shell had been thrown into the enemy's entrenchments
—the 68 pounder shots were seen to cut down large date
trees, 18 inches in diameter, as if they had been mere twigs—
and it was a matter of surprise, says Havelock's biographer,
that the Persians were able to stand the awful fire for so long
a period.   The signal was now hoisted for the steamers with
the troops to move up to the point of debarkation.   The
Berenice, carrying the 78th, with whom was Havelock himself, led the way.   The deck of the steamer was crowded, and
the General, glass in hand, took his position on the paddle-box.
He stationed himself there that he might the better watch
the movements of the enemy, and keep control of the men,
who were now passing entirely defenceless within a hundred
yards of the Persian guns.   A single round shot striking that
crowded deck would have played terrible havoc in the dense
mass of men who stood packed together.   As it was, the
guns were once or twice discharged at them, and the hull
and rigging of the "Berenice" were struck by several shots,
but not a life was lost.   The troops were landed under
the protection of the ships of war—Sir James Outram
arriving on the scene as the debarkation of the 78th was
being completed.

By half-past one the men were ready to advance.   Intervening
between them and the enemy's entrenched camp were a date
grove and a level plain, and through the grove and over the
plain went the force in splendid style, marching in contiguous
columns at quarter distance.   The position was formidable,
but the troops were on their mettle, and were determined to
carry it with a rush.   The enemy, however, gave them no
opportunity of displaying their fighting qualities.   The sight
of the armed force marching across the plain towards them
was too much for the poor Persians, who, frightened for their
safety, exploded their largest magazine, threw away their

weapons, and fled for their lives, leaving their baggage and other spoil to fall into the hands of the conquerors.

The Persians retreated to the town of Ahwaz, about 100 miles distant, on the river Karon ; and Outram immediately decided to send an armed force of 300 men, with small steamers, gunboats, and cutters to effect a reconnaisance. The command was given to Captain Hunt, of the 78th, who took with him two companies of that regiment. On the 1st of April they came upon the enemy, but instead of merely making a reconnaisance, as instructed, the gallant Hunt determined to engage the Persians. It was an adventurous enterprise, as the enemy had 6000 infantry, 5 guns, and about 2000 cavalry. But so pluckily did the Highlanders and the Grenadiers of the 64th, by whom they were accompanied, advance to the attack, so well were they disposed by Captain Hunt, and so courageously did they engage the Persians, that in an hour and a half the latter were in full retreat, leaving a large quantity of arms and stores, which was taken possession of by the plucky little expedition.

The Persians had now discovered that they were unable to meet British troops in battle with any chance of success, and peace was sued for and obtained. Fighting was thus at an end, and the force left the country.

Although there had been little opportunity of displaying their valour, the conduct of the 78th Highlanders had strongly impressed General Havelock, who was a close observer of men. In his confidential report on that corps, before leaving Persia, a copy of which, says Mr J. Clark Marsham, was found among his papers, the following passage occurred :—
" There is a fine spirit in the ranks of this regiment. I am given to understand that it behaved remarkably well in the affair at Kooshab, which took place before I reached the army ; and during the naval action on the Euphrates, and its

landing here, its steadiness, zeal, and activity under my own observation, were conspicuous. The men have been subjected in this service to a good deal of exposure, to extremes of climate, and have had heavy work to execute with their entrenching tools, in constructing redoubts, and making roads. They have been, while I have had the opportunity of watching them, most cheerful; and have never seemed to regret or complain of anything but that they had no further chance of meeting the enemy. I am convinced the regiment would be second to none in the service if its high military qualities were drawn forth. It is proud of its colours, its tartan, and its former achievements."

Nearer than any of them knew was the time when the military qualities of the 78th were to be drawn out. When Havelock reached Bombay, on his return from Persia, the first thing he heard was the "astounding intelligence" that the native army of India was in revolt. "This," he wrote, "is the most tremendous convulsion I have ever witnessed; the crisis is eventful."

Eventful, indeed, it was for England and for India; and, as the sequel proved, especially eventful for Henry Havelock and the gallant 78th, the heroic "Saviours of India."

Chapter VI.

THE INDIAN MUTINY.

THE annals of war have no darker page than that furnished by what General Havelock described as the "eventful crisis" which in May 1857 arose in India. It is the blackest chapter in British military history. India and Britain alike were startled and stunned by the suddenness of the rising. It came like a bolt from the blue, yet with the disastrous consequences of a mighty convulsion. In one day the work of Clive and Hastings, the Metcalfes and Lawrences seemed practically undone. Horror followed surprise, as the story of the mutineers' excesses was flashed from East to West, accompanied by the despairing cry for help of the sufferers in their extremity. It was a terrible time. The mutineers seemed suddenly transformed from men to fiends. They became possessed by a mad craving for vengeance, which the perpetration of no atrocity could satisfy. The more they revelled in bloodshed the more revolting became the forms of their fiendish cruelty—their hellish delights putting the direst acts of Nero himself to the blush. It was a revolution of the most overwhelming nature, and its details sent a thrill of indignation through the civilised world.

We shall endeavour, very briefly, to let the reader understand the situation. The administration of Indian affairs was then in the hands of the East India Company, which maintained at the time a large army, amounting to almost 250,000 men, in the main officered by Europeans. Of this army about 20,000 were Europeans—Queen's troops—but paid and maintained by the Company; the rest were native soldiers or Sepoys. The British regiments were spread through the Presidencies of Bengal, Madras, and Bombay, and were, compared with the numbers of the Natives, a comparatively slender force. Among the natives an idea had begun to prevail that in the hundredth year after Lord Clive's great victory at Plassey the British power in India would cease. For a time a growing discontent had been visible among the Mohammedans—no doubt smarting under the memory and oppression of the bitter wrongs which characterised our early rule in India—and this was increased by certain acts of the British authorities, which were considered to cast indignities and humiliation upon one or two of the great Native Princes. The signs of discontent were unheeded—the European force was indeed reduced, which gave the natives hope of deliverance—and the wave of disaffection spread from caste to caste, gaining in strength as it went. "The real motive of mutiny," says G. O. Trevelyan, writing, perhaps, a trifle too much from an Anglo-Indian point of view, "was the ambition of the soldiery. Spoilt, flattered, and idle, in the indolence of its presumed strength, that pampered army thought nothing too good for itself and nothing too formidable. High caste Brahmins all, proud as Lucifer, they deemed that to them of right belonged the treasures and the Empire of India. Hampered with debt, they looked for the day of a general spoliation. Chafing under restraint, they panted to indulge themselves in unbridled rapine and license." An emblem of agreement was secretly passed from station to station, and from hand to hand, by which each native regiment pledged itself to stand by its neighbour, each Sepoy to stand by his comrade, come what might. Experienced men saw the signs, and feared something was to happen; but of its nature they knew and dreamed nothing. And so Anglo-Indian society remained quite confident and careless—dipping in pleasure and gaiety, attending balls and conversaziones, marrying and giving in marriage, scheming and aspiring to place and power, and making engagements which would never be kept.

At length the spark was applied to the well-laid mine. It had been determined to issue the Enfield rifle to the Native troops. The cartridges of the Enfield were lubricated with grease, and those for the rifles now being issued were made in the great factory or laboratory at Dumdum, a few miles from Calcutta. One day a mechanic employed at the work asked a half-caste Brahmin for a draught of water out of his lotah—a small brass drinking-vessel. This the Brahmin refused, on the ground that he did not know to what caste the mechanic belonged—caste being to the Brahmin more precious than life itself.

"You are very particular about your caste now," sneered the workman in retort; "but you will soon have no caste left; for you will be required to bite cartridges smeared with the fat of pigs and cows."

To the Brahmin this intelligence was simply astounding. Against the idea his whole soul rose in horror and loathing. The Brahmin, whatever he may be in character or religious belief, is inexorable on the subject of caste. He must abide by his meal of milk, rice, or vegetables; to permit beef or pork to touch his lips would be a disgrace in this world and destruction in the world to come. The man, stricken frantic by what he had been told, flew from room to room, telling his comrades of the design the Feringhees had formed against their race. Hindoo and Mohammedan joined in the belief, and the story spread from station to station with the utmost rapidity.

Immediately the flame burst forth. At several stations the soldiers objected to the cartridges. They were dealt with in various ways; but still the disaffection spread. At length at Meerut the 3rd Native Cavalry refused to take the cartridges on parade, and next day nearly 90 of the men were tried by court-martial, then stripped of their uniform, ironed, and marched to prison. As they went they vowed vengeance against the Government and white-faced Feringhees. Two days later, Sunday evening, as the church bells were ringing, incendiary fires were seen to burst forth in various directions. Then the men rushed to the prison, burst open the doors, struck the chains from the prisoners, and let all within the walls go free. On the parade ground they shot dead Colonel Finnis and other officers, who endeavoured to appease them ; then, joined by an infuriated mob, they rushed to the houses of the European residents, and, without discrimination as to age or sex, butchered all they could seize, " aggravating murder by outrages still more horrible."

This was the opening of the flood of mutiny and bloodshed, and the Sepoys instantly let loose all the worst passions which savage and fanatical men possess. The cartridges they refused to take from the British officers they now freely used against them. From Meerut the mutineers went to Delhi, the ancient seat of the Mogul dynasty. It was attempted to shut the gates against them, but too late. They dashed into the city, and were joined by the troops there. Mr Fraser, the British Commissioner, was met and hewed down by their swords—his head being cut off and borne through the streets in triumph. At the palace gate they asked for Captain Douglas, the commander of the guard, who went and met them, but was shot dead before he could speak. The chaplain of the station they seized, and before the eyes of his daughter—an amiable and engaging young woman, about to be married—butchered him in cold blood, subsequently subjecting the poor girl to shocking indignities, then cutting her body to pieces. Sir Theophilus Metcalfe, the Governor of Delhi, saved his life by secreting himself in the city for three days, after which he concealed himself for ten in the jungle. Others were not so fortunate. A fell slaughter of Christian inhabitants took place. The insurgents plundered the Bank, and murdered Mr Beresford, the banker, his wife, and five children, by slowly sawing their throats with pieces of broken glass. In pillaging the premises of the *Delhi Gazette*, they hacked the printers to pieces. The British military authorities in the cantonments outside the city made an effort to quell the disturbance, but in vain. The Delhi men, by a preconcerted arrangement, suddenly rushed to one side of the road, leaving the officers on the other. Then they closed on the officers, so suddenly taken by surprise, and cut or shot them down. Colonel Ripley made a gallant fight, and shot two of the miscreants before he fell.

This was but the beginning of their atrocities. It soon became apparent to the British officers in the garrison that not a man among the Sepoys was to be depended on, and they made arrangements each to shift for himself. The bugle sounded the retire, and all who could made to flee from the city. A large party of officers and ladies were gathered near the Cashmere Gate, when the Sepoys suddenly appeared, and opened a tremendous fire on the helpless group. Shrieking and terror-stricken, all who were able of the poor women, some of them wounded and streaming with blood, fled and made their escape, most of them to die in the jungle, or to meet outrage, violence, and death at the hands of the surrounding villagers. In the city the inhuman monsters took, according to an officer, forty-eight females—" most of them girls from ten to fourteen, many delicately nurtured ladies, and kept them for the base purposes of the heads of the insurrection for a whole week. At the end of that time they made them strip themselves, and gave them up to the lowest of the people to abuse in broad daylight in the streets of Delhi. Then they commenced the work of torturing them to death, cutting off their breasts, fingers, and noses. One lady was three days in dying. They flayed the face of another lady, and made her walk naked through the streets."

In a mosque some Europeans had taken refuge, and after days of confinement, were almost beside themselves with the torture of thirst.

"Give us water," they cried to some Sepoys, "and we will yield ourselves to be taken before the King."

"Lay down your arms, and you will then get water," was the reply.

Their arms yielded up, the work of butchery began. Every one was put to death—eight officers, eight ladies, and eleven children—some of the latter being hung by the heels, and brained before their parents' eyes. At length the city was completely in the hands of the mutineers, the deposed King of Delhi was placed on his throne, and took the lead of the movement, being supported by native Princes in offices of responsibility and trust.

Still spreading, the revolt took the same form everywhere —white officers being shot down or bayoneted, and their wives and children tortured to death—all Europeans who could be seized subsequently suffering in the same way. At Lucknow, through the vigilance of Sir Henry Lawrence, one of the ablest and most clear-headed and politic of the Anglo-Indian military administrators, the mutiny was long delayed; but at length the men broke their allegiance, and, after some desultory fighting, left the city to join a general gathering of the mutineers at Nawabgunge, about eighteen miles distant. With the European garrison Sir Henry went forth to disperse the rebels, and although he did all that valour and good example could, he was overpowered by numbers, and, fighting every inch of the way, had to retire into the Residency, to the very entrance of which he was chased. British authority was at an end, and all that Sir Henry could do was to place the Residency in the best state of defence he could. Besides the soldiers—a small force—many helpless civilians, women, and children, and some professedly loyal Natives were hemmed in. They sustained a long and exhausting siege, enduring terrible sufferings. Provisions ran short, and Sir Henry again sallied forth, attacked the mutineers with 200 men, beat them back, and secured what he required. He was returning comparatively victorious, when the "friendly" Natives treacherously turned the guns they were using upon his party, and poured rounds of grape shot into their unsuspecting ranks. Unfortunately, Sir Henry Lawrence was himself wounded, and three days after died, leaving a band of devoted men and women to deeply mourn the loss of his heroic example and wise counsel.

At Bareilly the usual butchery took place ; but most of the women and children, and some of the officers, escaped, many of the latter exhibiting great courage in combating large numbers of the mutineers. At a place about fifty miles from Bareilly, the mutineers broke out while the officers were at church. They shot the clergyman as he ascended the pulpit. Lieutenant Spens was sabred as he knelt at prayer; the Doctor was shot as he drove up to church ; the Magistrate of the village was butchered in cold blood. The officers, with their families, escaped, to fall into the hands of treacherous guides, who had promised the fugitives every protection. On seeing Spens, who had not been killed, with his wounds bound up, they said it was no use taking with them a wounded man. Better shoot him at once, and shot he was directly.

Then they ordered the ladies to quit the carriage in which they were seated, and walk under the scorching sun. The officers remonstrated without effect—out the ladies had to go. On alighting they were shot one by one; some of the children were bayoneted, some dashed on the ground. The miscreants then killed all the officers, and subsequently the corpses were buried in a large hole dug in the ground. At Neemuch a Native officer persuaded the European officers to take refuge in an outhouse, when he turned a cannon upon them, and would have blown them to atoms, had not a Native, "true to his salt," secured their escape—all but a surgeon's wife, who, along with three children clinging to her skirts and shrieking for mercy, were cruelly butchered. At Benares a terrible struggle took place between the Native and British Forces, the Sepoys finally being repulsed. At Allahabad the duplicity of the 6th Regiment of the Native Bengal Army was conspicuous, even in that reeking hotbed of Sepoy treachery. The men had gone with tears in their eyes to the officers, beseeching them to trust in their fidelity, and so lulled suspicion. Some might have been in earnest; all were not so, and soon the effect of insidious lies and fanatical hatred became manifest. On the same night, at nine o'clock, the officers heard the bugles sound the alarm. As they hurried forth to ascertain what was wrong, fourteen of them were brutally massacred. Other officers met their death at different points. One poor fellow, who fell alive into their hands, was pinned to the earth by bayonets, and had a fire kindled on his body before death ended his sufferings. Then, joined by 3000 ruffians let loose from prison, the mutineers began their work of butchery. For the unfortunate Europeans in the town it was an awful night. Scores were killed in the streets, and many were subjected to the most fearful tortures. Their noses, ears, lips, and fingers were cut off; then their limbs were hacked, and death came as a welcome deliverer. Nor were the children spared—the poor little things being often dashed to pieces before their dying parents' eyes. At Jhansi the British officers had, suspecting danger, secured themselves and a number of women and children in a fort, which the mutineers attacked with cannon and musketry. After a heroic defence, during which the responsible officers were killed, the little garrison lost heart, and offered to lay down their arms if safety was promised. The promise was made, and kept as the others had been. Once outside the fort, the fugitives were surrounded, seized, and tied—the women being placed in one row, the men in the other. Then the massacre proceeded—the men being first killed, and the women and children afterwards. At Futteghur, where there were but thirty-three able-bodied Europeans, although with women and children the whites numbered a full hundred, a heroic defence was made, but at length the fugitives had to take to the river in boats, where they were pursued by Sepoys in immense numbers, and most of them put to the slaughter, others being taken captive to meet a still more horrible death in the charnel-house at Cawnpore.

It was at Cawnpore that ferocity and treachery revealed themselves in their most malignant form. Here the leader was the arch-fiend Nana Sahib—or Dhoondoo Punth, to give him his own name—a wretch filled with the most bitter hatred towards the Government and the British people. Cruel, crafty, self-willed, wealthy, and ambitious, this capricious son of indolence was accustomed to have his every wish gratified; and he hailed the mutiny as the means of enabling him to slip into much-coveted power. Early in June the tide of insurrection swept over the city; but the British, under Major-General Sir Hugh Wheeler, K.C.B.,

together with civilians, women, and children, had entrenched
themselves to the best of their ability, turning an hospital
barracks and the soldiers' church into temporary and feeble
forts. What they suffered no pen can describe. They had
to sustain for weeks a siege of the most determined kind.
They could get no rest, no sleep, and food was scarce. Subjected to incessant attacks under the burning Indian sun,
their strength was exhausted and their numbers reduced.
Around them and among them fell a constant hail of shot,
shrapnel, and bullets, and finally red-hot shot, fired on the
buildings they were so heroically defending, set their poor
shelter on fire. Nearly fifty of the weak and wounded
perished in the flames. And all the while no word of hope
from the outer world could reach the beleaguered brave.
They knew not that succour was approaching, and when the
treacherous Nana offered them, if they would lay down their
arms, free, unmolested passage to Allahabad, Sir Hugh, so
old and infirm that he should not have held so responsible a
post, consented, and the garrison came forth—300 women
and children, 150 soldiers, and an equal number of non-
combatants. For all the hideous details of the sequel we
refer the reader to Trevelyan's "Cawnpore." It is sufficient
for us to say that it had been no part of the plan that man,
woman, or child should escape. The mock arrangement was
that they should sail from Cawnpore in boats, and they
were escorted to the water by the rebel army, whose
demeanour to the wretched fugitives was anything but
encouraging. When they had embarked, a pre-arranged
signal was given, and the work of massacre commenced.
The surface of the water was swept by bullets, and heavy
guns threw their shot into the crowded boats. Bravely the
British struggled, but to no purpose ; soon the flowing
Ganges was red with blood, and filled with the dead and

the dying. Boats stuck on the banks, and the helpless occupants, moaning and shrieking, were butchered where they lay. One large boat, containing General Wheeler and party, got down the river a bit, but was recaptured, and all on board— 60 men, 25 ladies, and 4 children—taken prisoners, and reserved for the Nana's private slaughter-room. There the white-haired General, and all with him, were brutally shot— the brave women clinging to their husbands, from whom they refused to be separated. Miss Wheeler, the General's daughter, unhappily, was reserved for a worse fate. Still, after this butchery, there remained one hundred and twenty-two women and children, who were removed to the Assembly Rooms. A fortnight after, when Havelock and his avengers were at hand, under the immediate direction of the Begum—a creature of the Nana—a number of troopers entered the rooms where the prisoners lay huddled together, and the work of death commenced. This took place late in the evening; in the morning sixteen · poor wretches who had escaped the general carnage were hewn to pieces, and all the bodies were dragged forth and flung into an outside well, to rot and pollute the already foetid air.

We have but glanced at the outstanding incidents of the mutiny; of its myriads of details we have not spoken. Yet enough has been written to show the nature of the rising, and the atrocities by which it was accompanied. India was ablaze. In garrisons here and there over the country little bands of brave men, hampered by helpless women and children, held the hordes of the enemy at bay. Over their brave deeds we cannot linger; yet it is but just to say that, terrible as the convulsion was, and near as appeared the over-throw of British power, those gallant little garrisons, however closely beleaguered, fought like masters and conquerors, not like men beaten and despairing. Often the British officers.

taking action into their own hands, attacked and shot down
their rebellious men, by sheer force of their courage and bear-

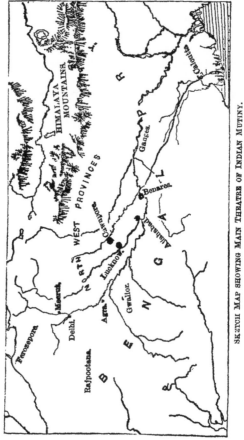

SKETCH MAP SHOWING MAIN THEATRE OF INDIAN MUTINY.

ing cowing the disaffected Sepoys into submission. And if
the men were brave, the women rivalled them in the quiet

courage they displayed, and in the readiness with which they did their utmost to aid those on whose strong arms and unwearied vigilance the safety and lives of all depended.

This was the crisis, then, in which General Havelock was so suddenly called upon to act. He was, as we have seen, at Bombay when he received the intelligence. To him its gravity was clearly apparent, and he immediately sailed for Calcutta. There he had been preceded by the 78th Highlanders, who were about to enter with all their strength into the terrible struggle.

Chapter VII.

THE INDIAN MUTINY—ADVANCE OF HAVELOCK'S COLUMN TO CAWNPORE.

WHEN Havelock, after enduring shipwreck and its attendant anxieties and dangers, reached Calcutta, it was to learn that British authority was extinct in the North-West Provinces, and practically so throughout the greater part of Bengal. He was called upon to take immediate action—to collect a force of British troops, and such Natives as could be thoroughly depended on, and advance immediately into the disaffected country. He was definitely instructed that, "after quieting all disturbances at Allahabad, he should not lose a moment in supporting Sir Henry Lawrence at Lucknow, and Sir Hugh Wheeler at Cawnpore; and that he should take prompt measures for dispersing and utterly destroying all mutineers and insurgents."

The task before Havelock was no light one; it was indeed one of the most arduous and trying in which military commander ever engaged. He had to march and fight, now under the sweltering Indian sun, and anon under drenching rains, with few men, without Cavalry, and with but an insignificant force of Artillery, all the guns being drawn by cattle. He could muster only 1,400 British bayonets—some of the men being armed with the old musket—and against him, barring every mile of the way, were the well-armed insurgents in their thousands. Besides, every man, woman, and child in town and village were enemies, and would do all in their power to harass his troops. Yet he entered upon his duty with the enthusiastic ardour of youth, rather than the halting hesitation of age. He had now attained the summit of his ambition. For the first time in his career he had placed under him a force of which he took supreme command—for the first time he entered upon a campaign the details of which were left to his own judgment, and the wielding of his materials to his own skill. He was proud of his position, and devoutly prayed for success. "May God," he wrote to his wife, "give me wisdom to fulfil the expectations of the Government."

The General went forward at once to Allahabad, which had already been practically subjugated by Colonel Neill, and there concentrated his Force. The column consisted "of

about a thousand bayonets, from four European regiments—
the 64th, the 78th Highlanders, the 84th Foot, and the 1st
Madras Fusiliers—with 130 of Lieut. Brayser's Sikhs, about
18 Volunteer Cavalry, and 6 guns." The 78th were,
as we know, fresh from the banks of the Euphrates, where
they had displayed the good promise which was now to bear
fruit in heroic performance. They were, however, badly clad
for the work before them. They had arrived at Calcutta,
says Mr John Clark Marshman, Havelock's biographer, with
the woollen clothing in which they had gone through the
Persian campaign, and no other, and no arrangements had
been made to furnish them with clothing suited to the month
of July. Havelock himself made every effort to supply them
with a lighter dress—fitted to a march with the thermometer
above 100 ; but in spite of every exertion, the army contractors
were not ready in time, and the 78th fought every battle of
the campaign in their heavy woollen kilted dress.

On the 7th of July—in the afternoon—the gallant little
band marched out of Allahabad, glared and scowled at by the
Natives as it defiled through the streets. Its destination was
Cawnpore, and its first experience was a drenching rain.
This compelled the men to halt for the night, after they had
marched for three or four hours, and got but a very little way
from Allahabad. Already the General had heard of the
murder of Sir Hugh Wheeler and his garrison, but he hoped
to find many Europeans still safe—if in extremity—and he
was anxious to get forward with all possible speed. Yet he
could not advance so quickly as his eager spirit desired. The
64th and 78th had been cooped up in steamers for weeks,
and although strong and robust men, were soft-footed and in
"bad form" for marching. The Madras Fusiliers were worse
—they had among them nearly 360 recruits, and the first
march found the road lined with fallen-out men. The end

of three days saw the column not much over twenty miles on its way. Yet there was urgent necessity to hurry on. The General learned that a small advance column which had been thrown forward, under Major Renaud, towards Cawnpore, was in danger of destruction from a large force of Nana Sahib's men which was moving against it. He therefore made forced marches to succour Renaud, and was successful in overtaking that officer before the enemy came up. It was at midnight on the 11th of July, in clear moonlight, that the Forces joined, and they continued to advance together till seven in the morning, when they reached the village of Belinda, about four miles short of Futtehpore.

After the severe exertions made by his troops, Havelock would gladly have given them rest on the 12th, but the General suddenly discovered that the mutineers were before him. They were moving hurriedly forward, to attack, as they supposed, Renaud's weak force. As the footsore and worn-out soldiers were beginning to cook something for breakfast, a couple of round shot suddenly ploughed up the ground in their midst; and as the men, leaving all thoughts of breakfast and cooking behind, sprang to their arms and formed up, the rebel horsemen, in frantic haste shouting and cheering, came dashing over the plain towards where they stood.

But they quickly reined up. Instead of the handful they had expected, they found a miniature army in perfect order waiting their approach. For a moment their hearts swelled with exultation as they gazed on the kilted Highlanders, and persuaded themselves that, the Feringhee men being all slain, their wives had now come to offer feeble fight. But the strange, self-reliant calm and order pervading the ranks soon filled the mutineers with forebodings. They would have retired as they came—overwhelming as their numbers were—

but the blood of the avengers was on fire, and through the British ranks coursed a wild desire for battle. It was shared by all—from the grey-haired General himself to the smallest drummer-boy; and nowhere did the spirit of retaliation burn deeper than in the breasts of the brave 78th Highlanders. Before the British column at length were the miscreants who had dyed their hands in the blood of the helpless and innocent—the cowardly butchers who had spared neither mother nor child. Sometimes soldiers fight mechanically, as automata move, having no interest in, and no knowledge of, the cause of quarrel. But in Havelock's column each man made the quarrel his own, and in his breast was a feeling of bitter, vengeful indignation against the foe. To-day it was neither crouching civilians, helpless women, nor clinging, crying babes that the Sepoys had before them to hack and maim; but armed men, strong in faith and hope, and eager for battle. "They had insulted my camp," wrote Havelock afterwards, "and their fate led them on to the retribution which awaited them." They would have run, but there was no escape. The General thought it best for the *morale* of his troops to let them go forward. It was with a grim feeling of pleasure that he saw in a portion of the enemy before him the 56th Native regiment—the very men he had commanded at Maharajpore—and through his mind passed the challenge—"There's some of you that have beheld me fighting; now try upon yourselves what you have seen in me."

And then the fight began. Strongly were the enemy posted, and splendidly were they armed; but they were unable to withstand the cool, steady advance of the resolute Europeans. Maude's Artillery was pushed forward, and the infantry supported them at long distance with the new Enfield rifles. The Sepoys could not stand the hot fire in which they were enveloped. First they were driven into the town, then through the town, and finally out of it and a mile beyond. Here they made a stand, and so exhausted were his men with the effort they had made, that Havelock despaired of being able to drive them further. But the spirit of vengeance was not yet dead, and it but needed the word from the old General to send the troops forward again with such dash and vigour that the enemy were in a few minutes wholly routed, leaving in the hands of the victors eleven pieces of cannon, some of which were a very valuable acquisition to the slender artillery of the British force. In the fighting the 78th had borne a conspicuous part, and had sustained themselves as Havelock expected they would. In the opening charge, their path lay through mud and water ankle deep ; but they followed their leader, Colonel Hamilton, in splendid style, as charging with levelled bayonets, they sent the enemy fleeing before them. In the subsequent movements they were well to the front, and so acquitted themselves throughout as to draw from the General in his despatch the remark that they were "full of spirit and devotion."

The victory gained, the tired, sorely worn troops sank down upon the ground they had so gallantly won. For twenty-four hours they had been marching, and had tasted no food since the preceding afternoon ; and now that they had for the first time since the commencement of the dreadful convulsion vindicated the supremacy of British arms, and turned, if ever so slightly, the tide of the rebellion's progress, their General thought them entitled to a day of repose. In the action the loss was merely nominal ; but the severe nature of the duty in which the men were engaged will be apparent when we state that during its progress no fewer than twelve died from sunstroke and fatigue. In his order of the day Have-

lock attributed the victory chiefly "to the blessing of
Almighty God on a most righteous cause," a novel form of
expression to address to British soldiers, but one which his
followers, esteeming him as they did, were not disposed to
ridicule, for they knew that if Havelock had great faith in
the God of Battles, so had he likewise trust in the stout
hearts and strong arms of his men. And beneath no red
tunic beat a braver heart than his own.

Early on the morning of the 15th the little Force, elated
with their victory, recommenced their march to Cawnpore.
They were now reduced in strength, as the Native Cavalry,
from whom treachery was feared, had been disbanded, and
100 Sikhs had been sent back to help to keep order in
Allahabad. The column had been but a few hours on the
march when, at the village of Aong, they found the enemy
again in front. This time the rebels were more audacious,
the Cavalry making numerous attempts to cut in on the
baggage guard, under the personal command of Havelock
himself, and plunder the baggage; but every assault was
repulsed, and finally the whole column moving forward in
extended order, the village was taken and the enemy driven
out. The want of Cavalry of course prevented the victory
being followed up. In this engagement Havelock lost the
valuable assistance of Major Renaud, who was shot in the
thigh while leading on his men. The gallant soldier died
three days after.

A five hours' march and a hard, if brief, battle gave the
troops a great appetite, and the halt being now sounded the
exhausted men set with alacrity about the preparation of
breakfast. They sorely needed the refreshment and stimula-
tion the meal would have afforded. But a sudden and
unexpected call was made upon their discipline and their
strength. A few miles distant, at the bottom of a gorge,

ran deep and strong the Pandoo river, now in high flood, and crossed only by a single bridge. The safety of this bridge had been a matter of anxiety to the General ever since he commenced his hazardous march. Once over it, he could, he thought, fight his way to Cawnpore. With it removed, he could not—poorly equipped as he was—cross the Pandoo in face of a strong enemy, and his expedition would thus suffer virtual defeat. And as his men set about the preparation of their much-needed breakfast, the report reached him that the enemy had retired to, and were rallying in strength at, the other side of the bridge, preparatory to blowing it up. His decision was taken in a moment. There could be no breakfast for his poor fellows that morning. Not an instant was to be lost—forward under that tropical, broiling sun they must go. He called on them to rise and advance. They did so at once, and without a murmur—convinced equally of the necessity of the case, and of the example of self-denial which their gallant leader was himself setting before them. "Those," says the General's biographer, " who know British soldiers can best estimate the value of the cheerful obedience they now displayed, as will always be the case when under a leader who has shown himself worthy of the men he commands."

Another two hours' march, and the gallant little relief column was again confronting the enemy. The rebels held a strong position, which they had been preparing for days. On the opposite side of the stream they were entrenched, and had planted a 24-pounder gun and a 24-pounder carronade, with which they could sweep the bridge and let nothing live upon it, or upon the Grand Trunk Road for more than a mile. But Havelock coolly began the battle, at the suggestion of Captain Maude, completely enveloping the bridge in a heavy artillery fire, under cover of which the infantry were pushed forward to shoot down the enemy's gunners. The fire was too hot for the enemy, who, frightened at the aspect of the stern white faces before them, attempted to blow up the bridge. Their purpose failed ; their train had been clumsily laid, and one only of the arches was damaged. The smoke had hardly cleared away when the Fusiliers bounded forward, and, before the gunners had recovered from their confusion, were upon them. The miscreants who remained to meet the furious onslaught were shot down or bayoneted where they stood, the rest took to ignoble, cowardly flight; the guns were captured—and the day was won ! The old General's promptitude had saved the bridge; had his advance been delayed an hour it would have been down, and the passage of the Pandoo would have been impracticable, except after a long, and to those waiting for succour, perhaps fatal delay. At two o'clock the now almost dead-beat troops, a mile on the Cawnpore side of the river, once more flung themselves on the ground to snatch a few hours of rest. Still misfortune was with them. The bridge was narrow ; the passage of the oxen and commissariat slow ; and by the time—late at night —the beasts could be killed for the men, the poor fellows were too exhausted to care for cooking. Many contented themselves with biscuit and porter, but the night being insufferably hot, in the morning when they would have eaten the meat they found it unfit for cooking, and had to throw it away and want.

Little did the gallant British soldiers know that they had just signed the death-warrant of over two hundred European women and children in Cawnpore. It was this night, as the tired and worn relief column rested on the banks of the Pandoo Nuddee—twenty-three miles distant—that the Nana Sahib completed his atrocities, and caused the massacre of the unfortunate women and children in the Cawnpore

Assembly Room, the bodies of whom were subsequently thrown into the famous well. He had learned of the passage of the Pandoo by the relieving force, and in the madness of his rage wreaked his vengeance on the innocents under his bolts and bars. Had the wearied soldiers but known, could they but have reached the scene in time, not a man in the broken ranks but would have pulled himself together and gone forward, through forests of the black demons, to the rescue of the helpless creatures, who, even then, were shrieking in their despair.

The morrow dawned, and the soldiers fell into their ranks and prepared to move onwards. Havelock was now beginning to feel the weight of the burden cast upon his shoulders. His force, slender at first, was by this time much thinned. Marching and fighting had cost him valuable lives. He was also short of ammunition, of commissariat stores, and of rum, which latter he considered almost indispensable in the circumstances in which his soldiers were placed. He was taxing their strength " to the utmost limit of human endurance, and he considered the aid of spirits necessary to sustain their physical powers." He had despatched messengers to Neill at Allahabad requesting him to send what he required—more men, stores, ammunition, and artillery; but he had to move on before they came. It was now the 16th of July, and not knowing that all the Europeans within Cawnpore had perished the night before, the column pushed forward heedless of fatigue. It was a day of excessive heat—the hottest which the men had yet experienced. "The rays of the sun," says a writer, "darted down as if they had been concentrated through a lens." Yet they toiled on, and had reduced the distance from Cawnpore by sixteen miles, when the village of Maharajpore was reached, and the order to halt was given. A rest of three

G

hours was allowed, and a meal—chiefly of biscuits and porter—was partaken of by the troops.

In the interval the General had not been idle. He had obtained information that the enemy were in force not more than a mile in front. With 5000 men the Nana had come out to meet him, and play his last stake for booty and power. He was supported by eight guns—five of them 24-pounders, and occupied a position of a very formidable nature. His left was covered by the Ganges and by four of the 24-pounders; his centre was posted in a low hamlet, with two heavy guns entrenched; and his right was behind a village, encompassed by mango trees, surrounded by a mud wall, and defended by two 9-pounders. The railroad embankment lay to the right of the position. The artillery was laid to sweep the road by which Cawnpore was approached. To attack this position, defended as it was by such numbers, was a task of no ordinary kind, and Havelock clearly understood its gravity. Yet the circumstances brooked of no delay; and once more his men were electrified by the kindling fire in his eye and by the order to stand to their arms. The General had resolved to stake all upon this daring attack. His plan of action was bold in the extreme; and nothing could have justified him in its adoption save the implicit confidence he had in his men, and his stern determination to strike strong and sure for victory.

The opportunity was now at hand, as he afterwards wrote, for which he had " long anxiously waited, for developing the prowess of the 78th Highlanders."

Chapter VIII.

THE INDIAN MUTINY—BATTLE OF CAWNPORE, AND SCENE IN THE CITY.

NANA SAHIB was no good soldier. He had made up his mind where Havelock was to strike, and how. He was to be attacked in front, he calculated, and he laid all his plans to meet a blow from that quarter. He saw none of the other moves on the board, and made no dispositions in view of them. But Havelock, with a true soldier's instinct, read his plans like an open book. He saw what the Nana expected, and what he did not; and resolved that the un-expected should happen to the rebel Force. So, instead of walking his column up to the face of the Nana's line, to be shot down by his strongly posted cannon, Havelock resolved to leave his baggage and wounded behind him, and try one bold, daring stroke for success. He decided to rapidly move his little Force to the left flank of the hostile line. Summoning his officers, he told them his plan, roughly sketching it on the road with the point of his scabbard. Then he ordered the troops to march, leaving baggage, camp-followers, and field hospital at Maharajpore.

Cheerily the men stepped out, although some of the 2nd Madras Fusiliers, having tasted no solid food for forty-eight hours, under the influence of their pint of porter, staggered as they went. Under cover of some mango trees, the column marched as rapidly as possible to the turning-point—the enemy meanwhile being apparently unaware of what was going on. Suddenly, however, through a gap in the trees, they spied the moving redcoats on their left, a sensation spread along their ranks, and with every gun that could be turned towards the flank of the 64th and 78th they opened fire on the British. Although men were falling, not a shot was fired in return—the column marching on with sloped arms, compactly and silently as if on parade, till they reached the point intended, and fairly turned the enemy's left. And now the whole movement was intelligible to the mutineers, for they made hurried dispositions to change their front and meet the coming attack. But they were already in the toils, and there was no escape. Quickly wheeling into line, the British column faced the foe, and advanced in echelon from the right, the 78th being the leading battalion. The Infantry were supported by the Artillery, which went forward as rapidly as the tired out cattle could drag the guns. The Artillery, however, had little effect on the half-concealed enemy, who continued to pour shot and shell with deadly effect into the advancing ranks. These guns had to be taken, and the General, turning to the Highlanders, who were still pressing on without firing, ordered them to undertake the perilous duty.

This was the opportunity for which General and men had been waiting. In a moment Colonel Hamilton, the grey-haired, brave old leader of the 78th, had ridden to the front and given them the word. Then the bagpipes' shrill pibroch pealed above the din of battle, and, giving a cheer which thrilled the hearts of all who heard it, the Highlanders followed their leader. Rapidly they sped forward, straight against the houses from behind which the three guns were vomiting forth their showers of grape. The Colonel's horse was shot under him, but the kilted soldiers never paused. Their formation was perfect, and with their arms at the slope, and marching steadily at the quick, they looked more as if engaged in a display parade than facing the storm and stress of actual warfare. But when within eighty yards of the muzzles of the death-dealing guns their aspect changed, they brought their bayonets to the charge, broke into the double, and like a pack of eager hounds dashed at the gunners. And now came a time of retribution to those black wretches who had the temerity to stand the shock of Gaelic fury. They went down like reeds before the terrible, crashing onslaught. In vain some of them dodged around and underneath their guns; those fearful bayonet points searched them out and pierced them, as they grovelled and shrieked for mercy to the avengers in whose raging bosoms mercy had been dried up. The guns were silenced, the village was taken, and the enemy, their left doubled up, were driven back upon their centre. This charge of the Highlanders was reckoned one of the most magnificent of the campaign, and Havelock wrote regarding it—"The opportunity had arrived, for which I have long anxiously waited, of developing the prowess of the 78th Highlanders. Three guns were strongly posted behind a lofty hamlet, well entrenched. I directed this regiment to advance, and never have I witnessed conduct more admirable. They were led by Colonel Hamilton, and followed him with surpassing steadiness and gallantry under a heavy fire."

The work of the regiment was not yet at an end. The Sepoys, falling back on the centre, had rallied behind the howitzer placed there. Meanwhile, the Highlanders, breathless by the rush and struggle in which they had engaged, had halted for a few minutes to rest and recover their formation. Five minutes had thus elapsed, when the General, who had followed close behind the first charge, rode up to the front of the line, and pointing to the masses rallying at the

howitzer, cried, "Well done, Highlanders! another charge
like that wins the day."

He was answered with a cheer and a shout, and emerging
from the shelter of a low bridge, behind which they had
re-formed, the 78th again dashed at the enemy, who, seeing
them coming with aspect furious as ever, and remembering
their experience of the previous conflict, fired a few stray
shots and fled, leaving the position to the conquering whites.

Away ran the mutineers Cawnpore-wards, over ploughed
and partially-flooded fields, in their speed outstripping
the tired avengers, who were unable to overtake them
in their flight. A village was before them, and towards
it they hurried, rallying again in face of the British force,
which, after a toilsome advance, had once more paused to
rest. From this shelter they opened fire, and Havelock's men
formed themselves again within range of their death-dealing
arms. At once Havelock, who seemed to be always at the
spot where he was most needed, hurried to the Highlanders,
and with high-pitched voice shouted—

"Come, who'll take that village; the Highlanders or the
64th?"

"There was no pause to answer," says Major North, who
narrates the incident. "The spirit of emulation was aflame
in every breast, kindled by the General's words. We (the
Highlanders) eager for approval, went off quickly in the
direction indicated, moving onward in a steady, compact line,
our front covered by the light company, and pushing the
enemy's skirmishers through the village, whence they were
compelled to fly."

And now the battle seemed won, and the rebel horde appeared to be in full retreat towards Cawnpore. Our men,
tired out with their terrible exertions under the broiling
sun, lay down to rest and light their cheroots, exchanging

hearty congratulations on the result of this their first general engagement and greatest triumph, and wondering whether the General would give them an extra allowance of rum. But their time for congratulations was short. Once more the mutineers had turned face. Nana Sahib, wild that this handful of white-faced warriors should rob him of the wealth and power which a few days ago he held as secure, had determined to make yet one last stand in defence of Cawnpore and all it contained which might gratify his passions and his greed. A heavy gun, planted on the Cawnpore road, suddenly opened fire, sending its shot with fatal precision into our exhausted ranks. Two bodies of horsemen galloped menacingly forward over the plain, bugles sounded, drums beat, a brilliantly-attired staff of officers, surrounding a still more gaily-dressed personage, announced that the Nana himself was present. For a time there was little stir in the British ranks. The worn-out cattle dragging the cannon were far away in the rear, and could not bring the Artillery forward. The eighteen volunteer horsemen (all the Cavalry that Havelock could boast) made such demonstration as was possible to check the enemy's advance. "The insurgents," says Trevelyan, "grew insolent; the soldiers were falling fast, and the British General perceived that the crisis was not yet over." He sent his aide-de-camp (his own son, now Sir Henry Havelock-Allan) to the spot where the men of the 64th were resting, and ordered them to rise and charge. The other regiments were ordered forward in turn, the 78th being next in echelon to the 64th. The enemy appeared in their thousands—the horse circling round the slender British force, now reduced to about 800 men, and the Infantry preparing to advance.

This, it was felt, was the crisis of the battle. There was no sign of weariness or fatigue, as the men sprang to their feet. "At the word forward," says an eye-witness, "the ardour and impetuosity of the troops rose to a height almost resembling frenzy." Young Havelock placed himself on horseback, and led the 64th straight for the thundering cannon on the roadway. Let Trevelyan describe what followed—"Then," he says, "the mutineers realised the change that a few weeks had wrought out in the nature of the task which they had selected and cut out for themselves. Now from left to right extended the unbroken line of white faces and red cloth and sparkling steel. In front of all the field officer stepped briskly out, doing his best to keep ahead of his people. There marched the captains, duly posted on the flank of their companies, and the subalterns gesticulating with their swords ; and the sober bearded sergeants, each behind his respective section. Embattled in their national order, and burning with more than their national lust of combat, on they came, the unconquerable British Infantry. The grape was flying thick and true. Files rolled over. Men stumbled and recovered themselves, and went on for a while, and then turned and hobbled to the rear. Closer and closer drew the measured tramp of feet ; and the heart of the foe died within him, and his fire grew hasty and ill-directed. As the last volley cut the air overhead, our soldiers raised a mighty shout, and rushed forward, each at his own pace. And then every rebel thought only of himself. Those nearest the place were the first to make away ; but throughout the host there was none who still aspired to stay within push of the British bayonets. Squadron after squadron, battalion upon battalion, these humbled Brahmins dropped their weapons, stripped off their packs, and spurred and ran, limped and scrambled back to the city that was to have been the chief and central abode of Sepoy domination. . . . At nightfall Dhondoo Punth (Nani Sahib) entered Cawnpore

upon a chestnut horse drenched in perspiration and with
bleeding flanks. A fresh access of terror soon dismissed him
again on his way to Bithoor, sore and weary, his head
swimming, and his chest heaving."

The British were victors here. The Cawnpore rebels were
crushed and dispersed. The first object of Havelock's advance had been accomplished. Few battles had been fought
in India which better than this displayed the splendid qualities
of British troops. " Fasting, footsore, and scarcely able to
bear the weight of their arms," and under a sun which was
as deadly as the enemy's grape-shot, they fought and conquered with a desperate valour which finds its counterpart
only in such glorious deeds as that accomplished by Wilson's
men in their terrible march of the 19th of January, 1885,
from the zereba through the Arab hordes to the Nile at
Metammeh.

The battle of Cawnpore cost over one hundred men, of
which the 78th had lost three killed and sixteen wounded.
The engagement was of short duration. A 78th man was time-
keeper. He was bugler to the General, and as the enemy's
first gun fired, Havelock gave him his watch to mark the
time. The bugler noted the moment before he put the
watch into his pocket, and as the last shot was fired after
the retreating enemy, he drew it out, and handing it to the
General, laconically remarked—" Two hours and forty-five
minutes, sir."

An incident of the engagement demands to be recorded.
During one of the advances of the British Infantry, a bugler
by mistake sounded the officers' call in rear of the 78th.
The officers of the 78th at once assembled near the General,
under the impression that he wished to speak to them. He
pointed out the mistake, but added—

"Gentlemen, I am glad of this opportunity of saying a few

words to you, which you may repeat to your men. I am now upwards of sixty years old; I have been forty years in the service; I have been engaged in action almost seven-and-twenty times; but in the whole of my career I have never seen any regiment behave better—nay, more, I have never seen any regiment behave so well—as the 78th Highlanders this day. I am proud of you, and if ever I have the good luck to be made a Major-General, the first thing I shall do will be to go to the Duke of Cambridge and request that when my turn arrives for a Colonelcy of a regiment I may have the 78th Highlanders. And this, gentlemen, you hear from a man who is not in the habit of saying more than he means. I am not a Highlander, but I wish I was one."

James Grant, in Cassell's "British Battles," narrates another incident of a diverting kind which is worth repeating. A corporal of the 78th Highlanders, he says, writing from Cawnpore to his friends in Glasgow, mentions that during the progress of one of the engagements it fell to be his duty, in company with another soldier, to carry a wounded Highland piper to the rear. They hoisted him shoulder-high, and were proceeding on their way when, to their astonishment, they saw a Sepoy on horseback riding furiously towards them with his sword drawn. The piper, who was wounded in the leg, raised himself up, and after going through the ordinary manœuvres of loading a gun, lifted the longest shank of his pipe to his shoulder, and pointed it to the Sepoy's head. No sooner had he done this than the valiant horseman turned tail and galloped rapidly away.

After the battle Havelock issued a general order, in which he said—

" Soldiers! Your General is satisfied, and more than satisfied, with you. He has never seen steadier or more elevated troops; but your labours are only beginning.

Between the 7th and 16th you have, under the Indian sun of July, marched 126 miles, and fought four actions. But your comrades at Lucknow are in peril; Agra is besieged; Delhi is still the focus of mutiny and rebellion. You must make great sacrifices if you would obtain great results. Three cities have to be saved; two strong places to be blockaded. Highlanders! it was my earnest desire to afford you the opportunities of showing how your predecessors conquered at Maida. You have not degenerated. Assaye was not won by a more silent, compact, and resolute charge than was the village near Jansemow on the 16th inst."

To his wife the General wrote :—"I am marching to relieve Lucknow. Trust in God, and pray for us. All India is up in arms against us, and everywhere around things are looking black."

The soldiers lay for the night on the ground outside Cawnpore, the General's waterproof serving him for a couch on the damp ground. The only food Havelock had tasted since the fight began was a bit of biscuit which happened to be in his son's pocket, and this he washed down with part of a bottle of porter which he got from Colonel Tytler. The night bivouac was grateful, but full of anxieties. The General slept with his bridle on his arm—his horse ready, saddled, standing behind, while his bugler lay at his side. No fires were lighted; no word was spoken; for the enemy's Cavalry still hovered round. During the night a deafening report broke on the ears of the British Force, then a cloud of smoke was seen to rise over the city, and they knew that the magazine had been fired by the rebels. It was their last act previous to flight; so spies reported, and so, too, did they report the full extent of the atrocities which had been committed on the helpless Europeans.

In the morning the troops entered Cawnpore, and then

they learned the worst. Every fond, lingering hope was
dashed to the ground. No European remained alive. They
found the building where the women and children had been
confined, and, says Trevelyan, "there was a spectacle which
might excuse much. Those who, straight from the contested
field, wandered sobbing through the rooms of the ladies'

MEMORIAL OVER WELL AT CAWNPORE.

house, saw what it were well could the outraged earth have
straightway hidden. The wives' apartment was ankle-deep
in blood. The plaster was scored with sword cuts ; not high
up as where men have fought, but low down and about the
corners, as if a creature had crouched to avoid the blow.

Strips of dresses, vainly tied round the handles of the doors, signified the contrivance to which feminine despair had resorted as a means of keeping out the murderers. Broken combs were there, and the frills of children's trousers, and torn cuffs and pinafores, and little round hats, and one or two shoes with burst latchets, and one or two daguerrotype cases with cracked glasses. An officer picked up a few curls, preserved in a bit of cardboard, and marked ' Ned's hair; with love ;' but around were strewn locks, some near a yard in length, dissevered, not as a keepsake, by quite other scissors. . . . There were found two slips of paper, one bearing in an unknown hand a brief but correct outline of our disasters. On the other a Miss Lindsay had kept an account of the killed and wounded in a single family. It runs thus, telling its own tale :—' Entered the barracks May 21st ; Cavalry left June 5th ; first shot fired June 6th ; Aunt Lily died June 17th ; Uncle Willie died June 18th ; left barracks June 27th ; George died June 27th ; Alice died July 9th ; Mamma died July 29th.' The writer, with her surviving sisters, perished in the final massacre."

MAUSOLEUM AT WELL OF CAWNPORE.

Horror-stricken, the soldiers proceeded in their search, and in crossing the outer courtyard perceived a number of human limbs protruding from a well. They hurried forward, and there saw a mass of bodies in every state of mutilation—the wounded having apparently been thrown in with the dead till the place was filled to the brim. The sight was one which none could witness unmoved. The men who had marched so bravely and so unmurmuringly from Allahabad, who had rushed to the cannon's mouth and faced the enemy's death-hail without flinching; and who had seen their comrades wounded, dying, and dead around them, now broke down and wept like "bearded babes." But sterner thoughts followed. Deep and bitter curses upon the miscreants who had done this deed burst from their heaving hearts, and vows of vengeance terrible to hear. They pulled the corpses from the well, and it is said, although we do not vouch for the accuracy of the statement, that when the Highlanders thought they had come upon Miss Wheeler's body, they cut the hair from her head, and sitting down and counting each man his portion, swore that for every hair a rebel should die.

Stern duties still lay before the gallant Relief Column. Havelock saw the storm gathering and thickening around, and he and his little band must needs go forth and brave its dangers. He had gloomy forebodings. The horrors he had witnessed depressed him, as he thought of the fate before himself and companions if overpowered in the struggle. But he had the faith and hope of a soldier. "If the worst comes to the worst," he remarked to his son, "we can but die with our swords in our hands."

Chapter IX.

INDIAN MUTINY—THE FIRST AND SECOND ADVANCES ON LUCKNOW.

WHEN Havelock had strengthened, as much as possible with the materials at his command, the defences of Cawnpore, he made preparations for an advance on Lucknow, the little garrison of which was bravely holding out against the furious attacks of the thousands of Sepoys by whom it was closely invested. The garrison was in urgent need of relief. By day or night it got no rest from thundering cannon and rattling musketry. Its head and counsellor, Sir Henry Lawrence, had died of wounds three weeks before, and brave Colonel Inglis had taken upon his shoulders the arduous duty of defending the Residency, and providing for the helpless hundreds dependent upon him. Incessantly the rebel assaults were continued, and day after day fewer of his comrades answered to the roll-call. By twos, and threes, and sixes the little garrison were daily shot down behind their barricades, and received into the hospital or the grave. Yet the survivors never flinched; stouter hearts never braved an enemy's fury, and more devoted women never ministered to suffering men.

Havelock was required to perform a race against time, and the question was—Could he win? On the 20th July General Neill came up to Cawnpore from Allahabad with such reinforcements as he could bring—only 227 men all told! Havelock then commenced his operations by crossing

the Ganges into the Province of Oude, which was one seething hotbed of insurrection. Getting his army over the
river, in face of an enemy of unknown strength, was one
of the most hazardous tasks ever undertaken by a commander
with so slender a Force. At any moment the enemy might
have descended upon him and played havoc among his
troops ; but, fortunately, the movement was effected in safety,
and by the 28th the relief column found itself at Mungulwar,
six miles beyond the river, with the enemy in immense
numbers and well-posted straight in front.

In Oude the Native army had been especially strong in
artillery, and had been trained to a high point of proficiency
in its use. The enemy that Havelock was now about to
engage was therefore strong in that arm in which he was
particularly weak. Nothing but a pressing sense of the
danger of delay for the Lucknow garrison would have in-
duced him, with so small a force, to attempt its relief ; but
believing in the justice of his cause and in the invincible
valour of his men—every one of whom he described as a
hero—he held with fixed resolution to his purpose.

It would be tedious to describe in detail this first attempt
(which was doomed to failure) to relieve the beleaguered
garrison, suffice it to say that the difficulties before the General
were found to be insurmountable. At Onao, three miles
beyond Mungulwar, where he first engaged the rebels, the
marshy, flooded state of the country rendered imperative an
attack in front against a position of unusual strength, and
defended with unusual tenacity. His troops carried the
position and captured fifteen guns, but these could not be
used for want of cattle to drag them. At Busseerutgunge,
six miles further on, to which he advanced after a three
hours' rest, he had a severe and stubborn fight, resulting in
victory, after many displays of genuine British courage.

These two battles had been fought in one day—with the exception of three hours' rest, the troops had been marching and fighting from sunrise to sunset. The men gloried in their double triumph, and a cheery evening was spent in camp after the day's fighting was done. But the morrow brought the General troublous thoughts. Before him he knew the mutineers were gathering in increasing strength, and on his left, during the progress of the previous day's fighting, he had seen a large body which had never been engaged. The victories, too, had cost him dear—eighty-eight men had been killed or wounded, and an equal number was dead or disabled by fatigue, exposure, and the ravages of cholera, which was carrying off the soldiers in scores. A third of his big gun ammunition had been expended in the two engagements, and he had not yet advanced a third of his way. Even were he allowed to proceed unmolested to Lucknow— which he would not be—there still lay before him a supreme effort which it was not in the power of his force to make. And, above all, he was concerned by the knowledge that every conveyance in which wounded could be placed was already occupied. With more wounded or sick men in his hands he could do nothing—save leave them by the wayside to destruction, and that he would never do. Painfully he realised his position, and regretfully took his resolution. He must have more men, and he must go back to get them— back to Mungulwar, where his position was so strong that he could hold the enemy at bay till he should be reinforced, and again able to go forward.

The march back, though dispiriting to the troops, was accomplished in safety, and Havelock then wrote to General Neill that he required another battery of Artillery and a thousand British bayonets " before he could do anything for the relief of Lucknow."

H

In the fighting which had taken place before the retreat, the 78th had borne an honourable part, and had added its quota to the killed. In the General's order of the day he says—" Lieutenant Bogle, 78th Highlanders, was severely wounded while leading the way at Onao into a loopholed house filled with desperate fanatics. A special report of his gallantry will be sent to His Royal Highness the Commander-in-Chief. The hero of Inkerman well knows how to appreciate heroes." Adjutant Macpherson, of the 78th, was also among those who were wounded.

Havelock had a further disappointment to sustain. The reinforcements which he had expected to receive were required elsewhere. The Dinapore regiments had mutinied, and a force of British troops, which Havelock might otherwise have obtained, had been sent to suppress the revolt. With the exception of about 250 men—scarcely sufficient to fill the gaps made by the sword and the ravages of disease— Havelock was told that he could get no further assistance for two months. This being the case, he felt that if Lucknow was to be relieved at all, the attempt must be made now. Every day would find his force less able for the duty, and each day's delay added to the danger of being too late in taking relief to the beleaguered garrison.

Once more, then, on the 4th of August, he moved out with his brave little band towards Lucknow. The first night he bivouacked a mile beyond Onao, the scene of a former conflict. Next morning he marched to Busseerutgunge, where for a second time the enemy was met. Havelock at once commenced the fight, sending detachments round to the rear of the town, and attacking with Artillery in front. After a fierce cannonade the enemy gave way, and, fleeing from the town, came under the fire of the turning force, which consisted of the Highlanders, Fusiliers, and Sikhs, with Maude's

battery and a troop of the Volunteer Cavalry. The fleeing
rebels did not pause till they were full five miles from the
town. And now the deplorable want of Cavalry in Havelock's
Force became apparent. Had he possessed Cavalry they could
have pursued the mutineers for miles, completed their
disorder, and cut them down in hundreds. As it was, with
no Cavalry following them up, the rebels, who had carried
off their guns, halted at Nawaubgunge, a spot chosen and
strengthened beforehand, and leisurely prepared to renew the
fight. The new position was reported by spies to be as
strong as the one from which the enemy had just been
driven, and beyond it, the General was informed, " the road
was dotted with posts equally difficult," where the enemy
would make determined stands. Besides, a bridge over the
river Sye at Bunee had been broken down, and the passage
was guarded by a large Force, supported by heavy Artillery.

Havelock was a brave man, and behind him he had the
soldiers whom he afterwards declared had been " the prop
and stay of British India in the time of her severest trial."
But the work before him he felt to be still beyond his
strength. If he went on, he would have to fight his way
through hordes of the mutineers, and would have to detach
men to hold each post he took. The population was hostile.
Every mile they would interpose and harass his troops, and
every village would have to be carried by force. He calculated that he would lose three hundred men at least before
he could reach Lucknow, still thirty miles distant, which
would leave him with seven hundred to attack " the city, with
its encircling canal, its entrenched and barricaded streets, its
loopholed houses, temples, and palaces, defended by a warlike
population, and an army of soldiers disciplined to perfection
by our own officers." Behind him were difficulties of a formidable nature. Nana Sahib, with 12,000 men, was at

Bithoor, and if the British advanced, would cut in and break his connection with Cawnpore. The Gwalior contingent—a splendid and compact force, consisting of Cavalry, Artillery, and Infantry—was also now in mutiny, and marching towards Culpee ; while the Dinapore regiments were hastening to join the Nana's standard. To add to the perplexities of the situation, cholera was again busy in the ranks of his little Force, and dozens were dying daily. Havelock had a strong desire to go forward. He knew with what longing hearts the besieged garrison would be waiting on his approach ; he felt that to accomplish his task would be an achievement worthy to be ranked amongst the most renowned deeds of the greatest of military heroes. He was very loth to turn back—a second time. But with all his desire to relieve Lucknow, he did not forget that he must have regard to the safety of his own men. Small and insignificant as his army was, it was the only body which represented British power and authority in the North-West Provinces of Bengal. If it were crushed, a new flame would be added to the raging conflagration of Sepoy mutiny, and the disaster might be irretrievable—involving, of course, the certain massacre of the beleaguered in Lucknow. He called his officers together, and told them that he could not sacrifice more lives in a hopeless advance ; they must go back—abandon the attempt—in the meantime, at least. More men were still wanted ; more men must be got. So back they trudged once more to Mungulwar, the men grumbling this time, for their hearts burned within them against the murderers, and they could not understand a retrograde movement when they had been on every occasion victorious. They did not see the thickening dangers all around ; they did not know all their General knew, all that he had to fear, when a few days later he wrote home to his wife :—" Things are in a most perilous state. If we succeed at last in restoring anything, it will be by God's *special* and *extraordinary* mercy. . . . I must now write as one whom you may never see more, for the chances of war are heavy at this crisis." He had still the same admiration for the Highlanders, and expressed it in the following terms, in a letter to General Neill, at the close of the second battle of Busseerutgunge—"If I might select for praise, without being invidious, I should say the Madras Fusiliers and the Highlanders are the most gallant troops in my little Force."

From Mungulwar he applied earnestly to Sir Patrick Grant, the Commander-in-Chief at Calcutta, for more troops, and while he waited he opened up communication with Neill at Cawnpore; finally, after immense exertion, succeeding in establishing connection by a bridge of boats thrown over the Ganges. He was now in a position to move either way —towards Lucknow or Cawnpore, whenever he had gained a sufficient accession of strength.

At this juncture Neill's small Force at Cawnpore was threatened by 4000 men and five guns from Bithoor. Neill applied to Havelock for aid, and this the General consented to give. But he could not attempt to cross the Ganges after a retrograde movement through Oude, else the report would spread among the mutineers that he had been chased out of the Province, and this, inspiring them with confidence, would be certain to involve an attack on his rear as he crossed the stream. He, therefore, immediately—on the 11th August— marched again to Busseerutgunge, and for a third time there engaged the enemy.

In this engagement the 78th Highlanders greatly added to their renown. They, along with the Fusiliers, the Sikhs, and Volunteer Cavalry, formed the right wing of the attacking party. The affair is so well described by Havelock's biographer that we quote his words:—"The right wing steadily advanced till their progress was suddenly arrested by an unforeseen obstacle. In front of the enemy's left lay a morass, covered with green vegetation, which presented the deceptive appearance of dry ground. The snare thus laid by the rebels was not discovered till the troops were on the verge of the swamp. They were immediately withdrawn, the 78th Highlanders moving on to the main road, while the Fusiliers, supported by four guns, passed round to the right. The enemy's guns were admirably served, and their fire was the severest our men had hitherto encountered. All the efforts of our Artillery, though superior in number, were unable for some time to make any impression on them, sheltered as they were by earthworks, and it was found necessary at length to have recourse to the bayonet. The Infantry of the enemy, posted behind the guns, continued to maintain a galling fire, but nothing could withstand the impetuosity of our troops. The Highlanders, now reduced in numbers to about a hundred, marched up to the guns, and when within a hundred yards of them, with their usual cheer, and aided by a flank movement of the Fusiliers, mastered them, and bayoneted the gunners. The Infantry then broke, and the Highlanders instantly turned the captured guns on them, and increased the confusion and slaughter. Our troops pursued them with unslackened energy through the town of Busseerutgunge, and over the causeway, which now for a third time became the scene of their defeat. The loss on our part amounted to 32, while that of the enemy fell little short of ten times that number."

In his order to the troops after this engagement the General said :—" The Fusiliers and the Highlanders were as usual distinguished. The Highlanders, without firing a shot, rushed with a cheer upon the enemy's redoubt, carried it, and captured two of the three guns with which it was armed.

If Colonel Hamilton can ascertain the officer, non-commissioned officer, or soldier who first entered this work, the Brigadier will recommend him for the Victoria Cross." Two brave young officers, Lieutenants Campbell and Crowe, were found equally entitled to the honour, having both entered the redoubt together in front of their men. Colonel Hamilton reported the difficulty in the circumstances of selecting either the one or the other. But the difficulty was too easily solved. Next day poor Campbell was stricken down by cholera, and Crowe, who survived, obtained the honour.

And now Havelock having struck once more home at the Oude rebels, and inflicted upon them a severe defeat, marched back to the Ganges, and on the same night crossed the stream, breaking up the bridge of boats behind him. Notwithstanding his troubles, sufferings, and perplexities, the General had not lost heart. Of a total force of 1,415 at Cawnpore, no fewer than 385 were down with sickness or wounds—10 men had died of cholera in one regiment in a single day. He foresaw that there was no choice between reinforcements being sent him and his Force being killed off by disease. But, meanwhile, fighting had to be done. The Bithoor Sepoys who threatened Cawnpore must be dealt with, and he at once marched against them. We need not detail the fight. It was " one of the strongest positions in India ;" and nowhere else had the Sepoys fought with such courage, even crossing bayonets with the British troops. But victory was not to be theirs. The British carried everything before them, and when they had conquered, cheered their General till he told them not to cheer him; it was themselves who had done it all. At night the exhausted troops were back in Cawnpore, and now had closed the first stage of Havelock's campaign to relieve Lucknow.

Chapter X.

THE INDIAN MUTINY—THE THIRD ADVANCE ON LUCKNOW.

HAVELOCK'S Relief Column had not relieved Lucknow, but it had given the mutineers a large experience of the quality of British soldiers. Seldom have European troops been called on to bear the fatigue of campaigning under circumstances so adverse. They had to march and fight either under the burning sun or amidst torrents of rain. They were often for twenty-four hours without food, and had never a bed upon which to lie down. Their only resting-place was the saturated or sun-baked ground, "carrying with them their sick and their wounded, and all their supplies, and suffering more from pestilence than from the weapons of the enemy." Then they had always to contend against a foe outnumbering them by about ten to one, as a rule better armed, and always strongly posted and prepared for battle. Yet they had been uniformly victorious —everywhere carrying defeat into the ranks, and dismay into the hearts of the rebels. Much of this success was due to superiority of race, and much of it to the enthusiasm with which the men espoused the cause; but to the brave old General who led them they owed a great proportion of their triumph. He was ever wakeful, watchful, and alert; ever careful of the lives of his men, ever mindful of their trials and sufferings, and ever skilful in the dispositions and arrangements which brought them conquest. He was brave, ardent, full of enthusiasm; and his presence always inspired his troops to the accomplishment of daring deeds.

It was, therefore, with feelings of no ordinary regret that they learned, during the time of harassing inaction while waiting for reinforcements in Cawnpore, that the General had been superseded in his appointment. Another officer higher in rank had been appointed to the command of the Relieving Force, and Havelock was placed under his direction. This was a mortifying blow to one who had so well vindicated Britain's honour, and so nobly sustained her cause. It looked like a mark of displeasure at Havelock's having failed to accomplish the relief of Lucknow, and but for the warm expressions of approval which were addressed to the General by those representing the Government in the matter, it would be difficult to understand it in any other light. It was, however, in many ways made evident that no slight was intended to Havelock, although no formal explanation was ever given. And if he was to be superseded, nothing could have given him greater satisfaction than to know his superior was to be none other than his old friend and comrade-in-arms, Sir James Outram. No two men could work better to each other's hands, and none knew better the exigencies of the kind of warfare in which they were engaged. That this union in office was to be mutually satisfactory the sequel will signally prove.

Havelock never for one hour forgot the object in view. Heartrending tidings had reached him, carried by heavily-bribed Native messengers, of the sufferings being endured by the beleaguered ones in Lucknow. Inglis, who was bravely holding the Residency, had been asked to cut his way out, if those coming to help could not cut their way in. But he replied that this was impossible. With his weak and shattered Forces he could not leave his defences. "You

must bear in mind," he wrote, " how I am hampered ; that I
have upwards of 120 sick and wounded, and at least 220
women and about 230 children, and no carriage of any
description. . . . In consequence of the news received,
I shall soon put this force on half rations. Our provisions
will thus last us till the end of September. If you hope to

GENERAL SIR J. E. W. INGLIS,

*The defender of Lucknow from the death of Sir H. Lawrence to the
Relief of the Residency by Havelock and Outram.*

save this Force, no time must be lost in pushing forward.
We are daily being attacked by the enemy, who are within
a few yards of our defences. Their mines have already
weakened our post. Their 18-pounders are within 150 yards
of some of our batteries, and from their position and from

our inability to form working parties we cannot reply to them, consequently the damage done is hourly very great. My strength in Europeans now is 350, and about 300 Natives, and the men are dreadfully harassed." Moved by this letter, Havelock urgently demanded reinforcements—which, at last, he was promised. Sir Colin Campbell had now reached Calcutta, as Commander-in-Chief of the army in India, and, thoroughly alive to the situation, exerted himself to send the sorely-needed troops. A letter from Outram reached the General, telling him that he should join him with reinforcements. This was the first letter Havelock had received from Sir James since his appointment over him in the command of the Relief Column, and in it Outram gave gratifying evidence of the chivalrous and generous spirit in which he intended to deal with the brave soldier whom he had superseded. "I shall join you with the reinforcements," he wrote; "but to you shall be left the glory of relieving Lucknow, for which you have already struggled so much. I shall accompany you only in my civil capacity as Commissioner, placing my military services at your disposal, if you please, serving under you as a volunteer." In no better way could Outram have displayed his confidence in Havelock, and his admiration for the heroic efforts he had made. Upon receiving this letter, Havelock sent an encouraging note forward to Inglis, and waited the coming of the fresh troops. This was on the 16th of August, and it was at dusk on the evening of the 15th of September that Sir James Outram reached Cawnpore, bringing with him about 2,000 men. Havelock was now also joined by Colonel Stisted of the 78th, and one or two companies of that regiment which had not hitherto been engaged.

The meeting between the two Generals was extremely affecting. Outram was the officer senior in rank, and had come as the Commander of the Force, but his first official act was to issue the following order :—

"The important duty of first relieving the garrison of Lucknow has been entrusted to Brigadier-General Havelock, C.B., and Major-General Outram feels that it is due to this distinguished officer, and the strenuous and noble exertions which he has already made to effect that object, that to him should accrue the honour of the achievement.   Major-General

Sir H. W. Stisted, K.C.B.,
Commanding the 78th Highlanders at first Relief of Lucknow.
Died Colonel of the 93rd Highlanders, 10th Dec. 1875.

Outram is confident that the great end for which General Havelock and his brave troops have so long and so gloriously fought will now, under the blessing of Providence, be accomplished.   The Major-General, therefore, in gratitude for, and admiration of, the brilliant deeds in arms achieved by General Havelock and his gallant troops, will cheerfully waive his rank on this occasion, and will accompany the Force to Lucknow in his civil capacity as Chief Commissioner of Oude, tendering his military services to General

Havelock as a volunteer.   On the Relief of Lucknow, the
Major-General will resume his position at the head of the
Force."

Havelock, whose own generous heart was brimming over
with gratitude and pleasure at this noble public act of
Outram, immediately issued the following order to the
troops :—

"Brigadier-General Havelock, in making known to the
Column the kind and generous determination of Major-
General Sir James Outram, K.C.B., to leave to it the task of
relieving Lucknow, and of rescuing its gallant and enduring
garrison, has only to express his hope that the troops will
strive, by their exemplary and gallant conduct in the field, to
justify the confidence thus reposed in them."

Sir Colin was gratified at this evidence, not only of good
understanding, but of warm personal esteem, between the two
Generals, and unhesitatingly confirmed the arrangement.   In
doing so he said—"Seldom, perhaps never, has it occurred
to a Commander-in-Chief to publish and confirm such an
order as the following one, proceeding from Major-General
Sir James Outram, K.C.B.   With such a reputation as
Major-General Sir James Outram has won for himself, he
can afford to share glory and honour with others ; but that
does not lessen the value of the sacrifice he has made with
such disinterested generosity in favour of Brigadier-General
Havelock."

This generous act of Outram has been much commented
on and much praised.   "It is certain," says one writer, "that
the moral dignity imparted by it to Sir James's character has
been incalculably more precious than any military renown he
could have acquired by the most brilliant success in command
of the expedition.   As long as the memorable events of the
Mutiny live on the page of history, and the memory of

Havelock continues to be cherished as a national heritage, this deed will be held in grateful remembrance." Another writer, Mr Headley, the American biographer of Havelock, refers to Outram's conduct in the following glowing terms :— "The days of chivalry can furnish no parallel to it. There is a grandeur in the very simplicity and frankness with which this self-sacrifice is made, while the act itself reveals a nobleness of character, a true greatness of soul, that wins our unbounded admiration. To waive his rank and move on as a spectator would have shown great self-denial and elicited the applause of the world; but not satisfied with this, he joined the Volunteer Cavalry, and though covered with well-earned laurels, stood ready to win his epaulettes over again. All his illustrious deeds in the field which have rendered his name immortal grow dim before the glory of this one act." It adds to our admiration of Outram, when we are informed that by this act of generosity he not only deprived himself of the honours to be won, but of the substantial reward which, as General Commanding, would have fallen to his share, and of the only means of support for the declining years of a life, the chequered vicissitudes of which had afforded no opportunity of making any provision for the requirements of age. It was an act of deliberate self-sacrifice.

We have dwelt thus fully on the incident between the two Generals, because it shows the kind of men under whom the gallant 78th were now so bravely fighting for their country. The heroism of the Highlanders had not only contributed to the distinction which Havelock had gained in the earlier struggles of this expedition, but to the confidence which Outram felt might safely be reposed in the men and the leader who had fought and won at Futtehpore, Jansemow, Cawnpore, Mungulwar, Busseerutgunge, and the other places where they had met and routed the enemy.

When Outram arrived, Havelock was ready to move forward. The Force was now to make its third—and as it proved—final and successful attempt to reach Lucknow. Inglis had written again, painting the situation more darkly than before. Day and night the rebels were incessant in their attacks, and the reduced rations were running done. No time was therefore to be lost, and every movement was made with urgent haste. Carrying fifteen days' provisions, the Relief Column found itself on the 19th of September once more across the Ganges, and marching on Mungulwar—the scene of former experiences. The rebels were there, but could not stand ; this time it was a resistless band of determined British soldiers, marching to succour their beleaguered brethren and almost despairing sisters, that came with resolute aspect to give them battle. The mutineers fled back to Busseerutgunge, and there again made show of fight. But on went the deliverers—their martial array, their glittering steel, and fluttering flags, the inspiring music of their bands, and the pealing notes of the Highlanders' pibroch, giving them now the aspect of men who had come to conquer. In vain the mutineers interposed their numbers and their fire. Through Busseerutgunge and over the Sye, in burning sun and pelting rain, went the British Force, driving the mutineers, in hot haste and confusion, towards Lucknow. Proudly did Havelock ride chief in the midst of his conquering band ; while Outram, grim, stern, and resolute, headed the Mounted Volunteers, leading them to victory whenever he gave his steed the rein, and winning, by his daring deeds, a Victoria Cross, which was never bestowed. On the 22nd of September the Column rested within sixteen miles of Lucknow, and a royal salute was fired, that its sound might reach the beleaguered garrison, and tell them of the help that was at hand. On the 23rd the march was resumed, and when the Alumbagh ("Garden of Lady Alum, or Beauty of the World") was reached, the mutineers were found waiting in order of battle.

Before the deliverers at last was the goal they had struggled so nobly to reach. Immediately in front was the stately palace of Alumbagh, set in a noble park, and surrounded by fair gardens, resplendent with the bloom of flowers and wealth of vegetation. Beyond, in the distance, glanced the towers and palaces of the city of Lucknow, brilliant with the ornate splendour of oriental magnificence. There, with anxious hearts, waited the weary ones, who for months had been doomed to a dreadful existence, fighting day and night to keep back the fiendish thousands who surrounded them and clamoured for their lives. Sick, diseased, famine-stricken, wounded, and worn, many of them lay, hope abandoned, and almost resigned to the fate which so many others had suffered—yearning for a sound from the outer world which would tell them that succour was at hand—for Havelock's royal salute of the day before had been unheard amid the clangour of their own hard conflict. And here, close at hand, was the advanced guard of the enemy to be overcome. An imposing sight the Sepoys presented. For two miles their line extended, and 12,000 mutineers stood ready to dispute the advance of the British Force towards the city. With them they had all arms—bayonet, sabre, and gun—and as the head of the British Column slowly moved forward along the trunk road their Artillery opened fire. But they could not stem the coming tide. Soon the British guns were thundering at their batteries, and the British bayonets levelled at their lines. They could not stand this resolute attack, and anon Outram and his horsemen were riding in pursuit, sabring the fleeing wretches as they sped towards the city.

The Alumbagh was taken, and many trophies were in the hands of the deliverers. Then, having possession of a base from which the final forward movement could be effectually made, the men lay down for the night, while the leaders concerted arrangements for entering Lucknow. The troops lay in advance of the Alumbagh, close indeed to the enemy's lines and batteries. The night passed quietly, but in the morning the guns of the mutineers opened furiously, the fire being directed to that part of the British line where the 78th were posted, and inflicting upon that already much reduced regiment a loss of 4 men killed and 11 wounded. On this day there was to be no engagement. Havelock had so determined. All, save a detachment of Highlanders under Major Haliburton, who took and held a stiffly-contested advanced post on the road to the Charbagh Bridge, were to reserve their energies for the work of the following day, and so, instead of being ordered, as they had so often been before, to silence the vomiting guns, the main body of the 78th, and the whole brigade to which they belonged, were simply commanded to retire to a place of safety. The day was spent in bringing forward the wounded, the baggage, tents, and camp followers, and placing them in the spacious rooms of the Alumbagh, which it was determined to hold—the command being given to Major M'Intyre, of the 78th. By afternoon the Alumbagh grounds had been transformed into a British camp, and the soldiers enjoyed their brief time of hard-earned repose.

Meanwhile Outram, who was best acquainted with the city and its approaches, was considering the plan of advance. He was not yet in command, but Havelock knew how and when to defer to his opinions and judgment. After mature deliberation, Sir James resolved on advancing by almost a straight course—they would force the Charbagh Bridge right

I

in front, and then fight their way up by the side of the
Canal, and through the streets to the Residency. This course
was agreed upon, and the plans of the leaders laid.

Another night passed quietly, and the morrow came—a
momentous day to those who had so long been waiting, a
day for which the whole civilised world had for months been
hoping—for all eyes had been bent on that solitary rock of
British heroism standing so bravely out amidst the seething
sea of disaffection. In the morning the troops were drawn
up. Havelock had spent some time in earnest prayer, and
"Patience" was given as the parole of the day. Outram, no
less earnest than his colleague, although giving less attention
to the outward and visible signs of piety, had spent most of
the morning carefully studying the map. As Havelock's
officers breakfasted together on a table in the open field,
Sir James joined them, and the final arrangements were
made. The soldiers had breakfasted comfortably—the last
meal many of them were ever to need—and now stood ready
for their orders. They were harassed and worn. "Toil,
privation, and exposure," wrote one who was present, "had left
traces on the forms of the men, yet daring, hope, and energy
seemed depicted in their countenances." Especially noticeable was the eager aspect of the men of the 78th, who trusted
to be appointed to the post of honour in the coming struggle.
Nor were they mistaken. Their leaders had the same confidence in them that they had in themselves. Between eight
and nine o'clock the order to advance was given—Outram
now taking the command of the leading brigade, which had
hardly moved beyond its own picket, when it was fiercely
assailed by the fire of the enemy.

Chapter XI.

UTRAM'S plan of attack, as briefly stated by his biographer, "was to force a passage across the Charbagh Bridge, thence to turn to the right and move along the bank of the Canal for nearly two miles, by the Bridge of Dilkoosha, on the road to the palace of that name; from which point a rectangular sweep of about three miles might be made to the Residency through the open ground east of the city."

At the Charbagh Bridge the first serious resistance was experienced. Here the Sepoys were strongly posted, and had determined to dispute the passage of the British. They had six heavy guns in position on the Lucknow side, one of them a 24-pounder, which completely swept the passage across the bridge, while numerous gardens and walled enclosures around were held in force. From this powerful battery and from the concealed Infantry a murderous fire was opened on the advancing column. Nothing could live under it. Outram, knowing of some high ground which covered the position held by the enemy, made a detour to the right, with the intention of operating on their flank and silencing their fire till the first attack on the bridge could be delivered. Meanwhile Maude's two heavy guns were dragged forward and opened upon the mutineers' battery; but they were unable to silence it. Almost every gunner was killed or wounded. For a time Maude himself and Maitland, his subaltern, remained unscathed, each pointing a gun, but so terrific was the enemy's fire that they had to call again and again for volunteers from the Infantry to fill the places of the fallen men, until it became apparent that this form of combat could not be successful, and that some other means must be taken to attain the object in view. Recourse was had to the bayonet. A desperate rush was made by the Madras Fusiliers. The first group who reached the bridge were shot down to a man. In that rush was young Havelock, and in a moment he was the only survivor on the bridge. He waved his sword, and called on the men to advance. They did so, and were on the mutineers before they had time to reload. At this moment Outram appeared, and the mutineers, caught between cross fires, fled, leaving their guns in possession of the British.

The bridge was gained, and now came the beginning of the struggle for the 78th. When the bridge was crossed our troops had practically entered the city. The direct road to the Residency—by which it was expected they would try to advance—having been cut by trenches, barricaded at short intervals, and rendered impassable, the British Force turned to their right, and pushed forward by the side of the Canal. The Highlanders were detailed to hold the bridge and the houses beyond until the troops, baggage, and wounded had crossed. Hastily the united column poured over, each regiment as it gained the other side following on after the one which preceded. On perceiving this movement, and that only a small body of troops held the bridge, the enemy returned in overwhelming numbers, and determinedly attacked the Highlanders. "Two companies, Nos. 7 and 8" (says Mr Keltie, who gives some details of the incidents of the day in which the 78th was concerned), "under Captains Hay and Hastings, were sent to occupy the more advanced buildings of the village ; four companies were sent out as skirmishers in the surrounding gardens, and the remainder, in reserve, were posted in the buildings near the bridge. The lane out of which the Force had marched was very narrow, and much cut up by the passage of the heavy guns, so that it was a work of great difficulty to convey the line of Commissariat carts and cattle along it, and in a few hours the 78th was separated from the main body by a distance of several miles. The enemy now brought down two guns to within 500 yards of the position of the 78th, and opened a very destructive fire of shot and shell upon the advanced companies, while the whole regiment was exposed to a heavy musketry fire. This becoming insupportable, it was determined to capture the guns at the point of the bayonet. The two advanced companies, under Captains Hay and Hastings and Lieutenants Webster and Swanson, formed on the road, and by a gallant charge up the street captured the first gun, which, being sent to the rear, was hurled into the Canal. In the meantime the skirmishing companies had been called in, and they, together with the reserve, advanced to the sup port of Nos. 7 and 8. The united regiment now pushed on towards the second gun, which was still annoying it from a more retired position. A second charge resulted in its capture, but as there was some difficulty in bringing it away, and it being necessary to retire immediately on the bridge to keep open the communications which were being threatened by the hosts who surrounded the regiment, the gun was spiked, and the 78th fell back upon the bridge, carrying with them numbers of wounded, and leaving many dead on the road. In the charge Lieutenant Swanson was severely wounded."

But the struggle was only beginning. Three hours had been passed in the hot engagement already described—the ammunition of the men had more than once given out, and required to be renewed. And now the entire line of troops, wounded, and carts having passed the bridge, the 78th formed up as the rearguard of the advancing column, and evacuated the position they had held. It was immediately seized by the enemy, who lined the right bank of the canal, and with a gun placed on the bridge and commanding the road, poured from all sides a terrific fire into the ranks of the devoted Highlanders. The 78th were almost completely surrounded, and every few paces had to turn and fight. Captain Hastings, at the head of No. 8 Company, made a determined stand against an advancing mass of the rebels, and was severely wounded. Almost at the same moment Captain Lockhart and a number of men were also wounded and disabled. Word had been sent forward to the main body that the situation of Stisted and the 78th was desperate, and some Cavalry and a company of the 90th Regiment were sent back to the High-landers' assistance. Still fighting every step, the now slightly reinforced men at length reached Major Banks' house. Here they lost their way. The main body, pushing on to the Residency, were far ahead, leaving no trace of the way they had gone. Turning to their left, instead of going straight on, the Highlanders took a different and more direct road than that by which the main body was proceeding. As they marched along they were exposed to a sweeping fire from loopholed houses in the streets, while the rebel horde, following up, harassed their rear. Presently they reached the Kaiser-bagh, or King's Palace, wherein was placed a strong battery, which was firing upon the main body, then fighting its way on the right in the vicinity of the Motee Mahul. The Highlanders came upon the battery in reverse, and with one rush captured the guns, then spiked them, and played havoc among the gunners. As they marched on, past the walls of the

Palace, they were subjected to a storm of fire from concealed rebels lining the whole length of its walls. Many a brave man fell without the possibility of his death being avenged, for the rebels were firing from secure hiding-places. The men were by this time terribly exhausted. They had been marching and fighting in isolated positions since nine in the morning, and now the afternoon shadows were beginning to lengthen. Finally, about four o'clock, to their joy they saw the main body before them, at the point where the different routes they had taken converged, and in the heart of Lucknow rose a ringing British cheer. The two bodies joined, and near the Fureedh Buksh indulged in a brief period of rest.

The Relief Column was still half a mile from the Residency, and every inch of the way had to be fought. It was growing dusk by the time the men were ready to move on again, and Outram, who had been wounded early in the day, counselled a halt at the Palace till morning. He wanted up the guns, baggage, wounded men, and rearguard before they pushed on. In the morning he thought communication might be opened with the Residency through intervening Palaces, "less brilliantly, perhaps," says Sir F. G. Goldsmid, "but with less exposure of life than by the street." But Havelock pressed to go on at once, and urged his authority respectfully but firmly to Sir James. He pointed out the straits to which the inmates of the Residency were reduced, the danger to which they were exposed from mines which might be sprung at any moment, the possible desertion of Natives under disappointment at the non-appearance of the deliverers, and their own want of strength to sustain a renewed and determined attack of the mutineers, should such be attempted. Then, not the least of the inducements to go on was the fact that not half a mile distant were the anxious hearts for whose sake he had been pressing and struggling for the previous three months. Outram yielded readily—not only yielded, but frankly and bravely offered to lead the way.

And now came the last terrible struggle. " The ever ready and gallant Highlanders," led by Col. Stisted, were, along with the Sikhs, summoned to the front. Then, Sir James and Havelock placing themselves at their head, the order to advance was given. They entered a very storm of fire and shot. The dusk was lit up by the incessant discharges of the enemy's muskets. In front, from every house, from every cross lane or street on their route, the advancing men were assailed. At point-blank range the mutineers poured in their shot, and brave men fell at every step. General Neill, bringing up the rearguard, had been shot through the head at the corner of a street ; young Lieutenant Kirby of the 78th, received his death wound as he proudly bore aloft and waved the Queen's colours in the face of the foe. But, thanks to Sergeant Reid and Surgeon M'Master, the flag was secured and carried into the Residency. It was a fearful struggle, but the deliverers were not to be stayed ; wildly the bagpipes screamed their slogan into the ears of the beleaguered brave within the goal, and valorously the blood-stained mountaineers fought and conquered. Outram, pale and bleeding, but with blazing eyes and resolute mien, rode on through the storm, sad at heart withal, for he was riding to the home, now a mass of ruins, where he and Lady Outram had spent many happy days. By his side was Havelock, calm, self-possessed, but thoughtful, for he knew that far in the rear his wounded son was being carried on a litter through the stress of the battle. Near by rode Colonel Stisted, of the 78th, watching and guiding his regiment through the fearful *melee*. Ever and again deep and artfully dug trenches crossing the street caused delay in progress,

while the mutineers, taking advantage of the stoppage, from
points of vantage sent volley after volley into the ranks of
the Highlanders and Sikhs. Within two hundred yards of
the Residency gate Lieutenant Webster was shot dead; but
a wild hurrah and a more desperate charge was the Highlanders' reply. Crowe and Macpherson and Jolly and Grant
dropped wounded—Macpherson for the second time in the
campaign—but others took their places, and on to the
Residency went the column, leaving behind its crimson trail
of wounded and dying.

Who can describe the feelings of those within the barricades? Two days ago they had been at the very point of
despair. Mine after mine had been sprung, and their weak
defences had been almost blown down about their ears.
Help had been so often expected and had never come, that
they had ceased to rely on it, and were almost counting the
hours when they would have to succumb to the persistent
and overwhelming attacks to which they were exposed. But
on the 22nd their drooping spirits had been suddenly roused
by the reports of spies that Havelock and Outram were at
hand. They could hardly believe the news, but on the day
following the thundering of cannon outside the city told
them it was true. What hours of suspense followed, as the
sound of firing came nearer and nearer! They could see the
commotion in the city, as the rebels prepared for defence
against the coming troops. Then followed the inactivity of
the 24th, and they could not learn what it meant. With
hands clasped, some of them sat and waited, bearing the tension as best they could. They could not eat; they could
not sleep. What a weary day and night—longer and more
trying than any they had yet endured. But with the 25th
had come the renewal of the battle, and they knew their
deliverers were approaching. Louder grew the din; nearer

and nearer came the sounds of conflict. All day the battle raged, and then towards evening every ear was electrified by the wailing notes of the bagpipes screaming above the rattle of musketry. The Highlanders are coming. Hurrah! And the wild cry was echoed from without. Hurrah, hurrah! uprose the British cheer from the lips of the advancing heroes, as every volley cleared their way and every step brought them nearer.

At last they were at the very gates. Wide open these were flung, and in poured the deliverers. It was now dark, but no matter. Brave men could shed tears the more freely; the possessors of husky voices would be less readily detected as they uttered their broken cheers. A staff-officer, one of the beleaguered, thus describes the scene :— " From every pit, trench, and battery—from behind the sand bags piled on shattered houses—from every post still held by a few gallant spirits rose cheer on cheer—even from the hospital many of the wounded crawled forth to join in that glad shout of welcome to those who had so bravely come to our assistance. It was a moment never to be forgotten. The delight of the ever gallant Highlanders, who had fought twelve battles to enjoy that moment of ecstacy, and in the last four days had lost a *third* of their number, seemed to know no bounds. The General and Sir James Outram had entered Dr Fayrer's house, and the ladies in the garrison and their children crowded with intense excitement into the porch to see their deliverers. The Highlanders rushed forward, the rough-bearded warriors, and shook ladies by the hand with loud and repeated congratulations. They took the children up in their arms, and, fondly caressing them, passed them from one to another in turn. Then, when the first burst of enthusiasm was over, they mournfully turned to speak among themselves of the heavy losses they had sustained, and to enquire the names of

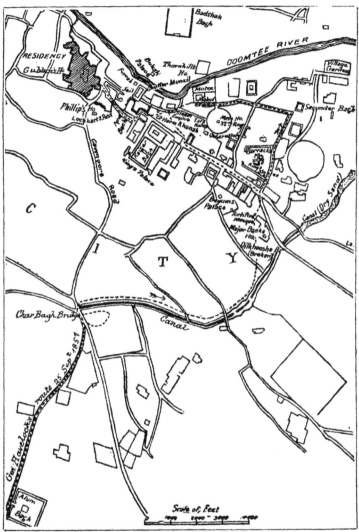

SKETCH PLAN SHOWING ROUTE OF HAVELOCK'S ADVANCE TO THE RESIDENCY AT LUCKNOW.

the numerous comrades who had fallen in the way." Another
of the defenders of the Residency, Mr L. E. R. Rees, also
writes of the joyous meeting :—" I was nigh bursting with
joy. The tears started involuntarily into my eyes, and I
felt—no ! it is impossible to describe in words the mingled
feeling of hope and pleasure that came over me. The
criminal condemned to death, and, just when he is about to
be launched into eternity, is reprieved and pardoned, or the
shipwrecked sailor, whose hold on the wreck is relaxing, and
is suddenly rescued, can alone form an adequate idea of our
feelings. We felt not only happy—happy beyond imagina-
tion—and grateful to that God of Mercy who, by our noble
deliverers, Generals Havelock and Outram and their gallant
troops, had thus snatched us from imminent death ; but we
also felt proud of the defence we had made, and the success
with which, with such fearful odds to contend against, we
had preserved not only our own lives, but the honour and
lives of the women and children entrusted to our keeping.
As our deliverers poured in, they continued to greet us with
loud hurrahs ; and as each garrison heard it, we sent up one
fearful shout to Heaven. . . . It was the first rallying
cry of a despairing host. Thank God, we then gazed upon
new faces of our countrymen. We ran up to them, officers
and men without distinction, and shook them by the hands,
how cordially who can describe ? The shrill notes of the
bagpipes pierced our ears. Not the most beautiful music
ever was more welcome, more joy bringing. And these
brave men themselves, many of them bloody and exhausted, forgot the loss of their comrades, the pain of their
wounds, the fatigue of overcoming the fearful obstacles
they had combated for our sakes, in the pleasure of having
accomplished our relief. How eagerly we listened to their
stories ; with what sentiments of gratitude, and pride, and

pleasure we heard what sympathy our isolated position had excited."

Three especial cheers were given for the leaders of the gallant band, and Havelock tells how the "half-famished garrison" contrived to regale him, "not only with beef cutlets, but with mock turtle soup and champagne." By this noble deed—the climax of all their triumphs—the 78th gained a glory that is imperishable while the page of history lasts. "Never," says Havelock's biographer, "did the valour of this gallant regiment shine brighter than in this bloody conflict." Every man fought for a Victoria Cross, but every man could not get the trophy. Yet the reward was due, and heroes, of whom more anon, were elected by the regiment to receive the representative token of valour. Both Generals were loud in their praise of the conduct of the Highlanders, and both specially mentioned them in their official despatches.

Yet this brave feat of arms "was no literal or effectual relief" to the garrison of Lucknow. It had been the intention of the Government, and of Sir James Outram, to at once withdraw the garrison and the sick and wounded to Cawnpore. With this view, the greater proportion of the baggage, ammunition, and provisions of the Force had been left behind at the Alumbagh, and the troops had gone on with scarcely anything but three days' food and what they had on their backs. It was now found, however, that the step was impracticable. The enemy numbered from 40,000 to 50,000 men, and were constantly increasing; the British, 2,000 strong when the attack commenced, had lost one-fourth of their number. The women and children found in the Residency amounted to 700, and the sick and wounded to 500. No carriages could be procured, and without these removal was impossible. The relievers would themselves require to be relieved.

Chapter XII.

IN the operations just closed the 78th had lost heavily;
but there were still arduous duties before the regiment.
The wounded and heavy guns had been left during the
night with the 90th at the Motee Mahul, and next morning a
draft of the 78th was sent to assist in bringing them in;
another draft was sent to keep open communication between
the Motee Mahul and the Residency. All day the post was
subjected by the rebels to a heavy cannonade, and the task
of conveying the wounded to the Residency was most difficult
and trying. During its progress a disastrous accident occurred.
A convoy, under charge of Surgeon Home, of the 90th, was
misled by a civilian, Mr Bensley Thornhill, the husband of
Havelock's niece, who, having gone out to bring in young
Havelock, had kindly offered to show the way back. They
suddenly entered a square filled with rebels, who immediately
surrounded the convoy. The doolie bearers fled, and the
escort, with a few wounded officers and men, took refuge in a
house. They were immediately besieged by from 500 to
1,000 of the enemy, against whom they defended themselves
with the utmost heroism. But many of the wounded fell
into the enemy's hands, and thirty or forty of them were put
to death—some, it is alleged, being burnt alive. Poor Lieut
Swanson, of the 78th, who had been wounded while bravely
fighting in the terrible struggle at the Charbagh Bridge, was
among the saved, but not until he had received two additional

wounds, one of which unfortunately proved mortal.   Among
the gallant men who specially distinguished themselves in
this unequal battle were Privates James Halliwell, Richard
Baker, and William Peddington, of the 78th; and the first
named and Surgeon Home received the Victoria Cross for
the valour they displayed.   Lieutenant Havelock was one of
the wounded, and, along with a wounded man of the 78th,
was heroically defended by burly Henry Ward, a private of
the Buffs, who stood by the side of the doolie under an awful
fire ; and although assailed by scores of the enemy, the brave
fellow fought like a tiger, and prevented the frightened
bearers from relinquishing their load.   Ward also, as he well
deserved, received the Victoria Cross " for this act of intrepid
gallantry."   The affair must, however, have ended disastrously but for the arrival on the scene of the main body
of the 78th, who had been detailed to a special duty, which
they had bravely accomplished, and, in returning to the
Residency, fortunately entered the square when the unequal
struggle was at its worst.   The rebels, when faced by the
new comers, quickly fled, and the besieged convoy was
rescued.   Bensley Thornhill expiated his blunder with his
life, dying. two or three days after from the wounds he
had received.

Of the 18 officers and 428 men of the 78th who left the
Alumbagh on the morning of the 25th of September, 122
were killed or wounded by the following day—2 officers
and 39 men being killed, and 8 officers and 73 men
wounded.   But the regiment had gained much in honour.
Their fame was destined to spread over India, and their name
to be regarded with pride and held in reverence at home.
Of the Victoria Crosses we mentioned in our last chapter,
one was awarded to the Adjutant of the 78th, Lieutenant
Herbert Macpherson, who had been wounded during the

second advance to Lucknow, but who, on the 25th of September, had sufficiently recovered, in the language of the official record, to " set an example of heroic gallantry to the men of the regiment, at the period of the action in which they captured two brass 9-pounders at the point of the bayonet." He was again wounded during the final advance to the Residency. Since that time, we may mention in passing, the then young Lieutenant has had an honourable and brilliant military career. In the Egyptian war of 1882 he commanded the Indian Contingent, and on the day on which these lines were written, the daily papers contained the announcement that " Major-General Sir Herbert Macpherson, V.C., K.C.B., has been selected as the successor of Major-General the Duke of Connaught in the command of the Meerut division of the Bengal Army."* A proud position worthily won, and one in which his old comrades of the 78th could not fail to accord him their heartiest good wishes. A regimental Victoria Cross was also awarded, and upon the vote being taken, this was unanimously bestowed upon Assistant-Surgeon M'Master, for the intrepidity he had displayed in exposing himself to the fire of the enemy when attending to the wounded on the 25th. His plucky conduct in carrying the Queen's colours to the Residency gate also evoked the warm admiration of the regiment. A Victoria Cross was likewise bestowed upon Colour-Sergeant Macpherson.

The death of Brigadier-General Neill led to the appointment of Colonel Stisted, of the 78th, to the command of the first Brigade ; and Major Haliburton, a brave and capable officer, who had done some excellent work during the relieving operations, succeeded to the command of the Highlanders. Poor fellow, he did not long enjoy this honourable position. A strong battery in Phillip's garden, close to the Residency, was

* See Appendix D.

a constant source of annoyance and danger to the garrison. This battery it was determined to capture or silence, and with this object Colonel Napier left the Residency with 568 men of all ranks, of which the 78th, under Haliburton, formed a strong contingent. The work was long and arduous. For two days the fighting lasted, but finally the enemy were driven off, the guns captured, and Phillip's house destroyed. Then, at the head of the sallying party, Haliburton for the next two days attempted to work his way from house to house with crowbar and pickaxe, with a view to open communication between the Cawnpore Road and the Alumbagh. During this proceeding hard fighting had to be done, for the mutineers were in a dogged, determined mood, and on the 4th of October poor Haliburton fell mortally wounded. "Where are you hurt, sir?" asked the surgeon of the dying officer. "In the front," was the proud reply, gasped out with failing breath. Almost at the same moment Haliburton's successor fell disabled, and the troops had to be withdrawn.

After this the arduous services of the 78th did not partake of that brilliant character which we have during the last few chapters described, and we do not propose to detail them at length. The regiment was much reduced in officers and men, and was divided into four posts, which, during the occupation of the Residency by Outram, who was now in supreme command, it continued to hold with unceasing bravery against incessant attacks, until the arrival of Sir Colin Campbell in sufficient strength to carry off the women, children, wounded, and treasure, and permit the Residency to be evacuated. In this difficult but entirely successful operation, to which we shall have another opportunity of referring, the 78th was allotted the post of honour. It formed the rearguard, and Outram rode with

K

it. He told the men he had selected the 78th for the honour of covering the retirement of the Force, as they had had the post of honour in advancing to relieve the garrison, and that none were more worthy of the distinction. On the 22nd of November the Residency was evacuated, and the British Force retired to the Alumbagh. Here a sad event soon after occurred. General Havelock, harassed and worn, sickened and died, and the heroes of the 78th wept bitter tears over their loss. They had learned to love him, and his death, coming soon after the signal success of their united efforts, and ere he had reaped the full rewards of his triumphs, made his end, to them, peculiarly saddening. He was buried in the grounds of the Alumbagh. Sir Colin Campbell carried off to Cawnpore the women, children, and wounded of the garrison, but Outram was placed in position at the Alumbagh and adjacent posts, with instructions to hold them until Sir Colin—who had in the meantime to retake Cawnpore and reconquer the Doab—could return in sufficient strength to finally attack and reduce Lucknow.

In this campaign the 78th had been admirably officered. Hamilton, Stisted, M'Intyre (who had held the Alumbagh for six weeks), Lockhart, and Bouverie were especially thanked for their services—M'Intyre and Lockhart subsequently in turn reaching the Lieutenant-Colonelcy of the regiment. A Victoria Cross fell to the lot of Surgeon Jee, who had throughout the whole campaign, and under the most trying circumstances, acted with great skill, unwearied patience, and indomitable courage. In the final siege of Lucknow the Buffs, reduced, jaded, and worn, were not a part of the actively aggressive force, and had no opportunity to display the high qualities they possessed; nor had they many opportunities of attaining signal distinction in their subsequent services during the operations of the Rohilcund Field

Force. At Bareilly, however, when they did get a chance, they gave the Ghazies emphatic evidence of the stuff they were made of. The affair at Bareilly practically ended their share in a campaign which had tried them as few regiments have been tried in modern times.

Between Sir James Outram and the 78th Highlanders a warm feeling of mutual esteem existed. They had occasion to admire the high qualities of the General, both as a soldier and a man, during that campaign on the Euphrates where they had served together. And he, quick and observant, understood what manner of men they were, and took no pains to conceal the good opinion he had formed. An instance of this mutual admiration, related in Major-General Goldsmid's Life of Outram, may be quoted. At the conclusion of the Persian campaign the regiment had been inspected by Outram and Havelock, before leaving Mohumrah for India. The inspection had taken place in the morning; but the men would not leave without taking a special farewell of their gallant chief. They had intimated their wish through their officers, and Colonel Stisted had arranged with one of Sir James's staff that the General should be detained in his tent to receive them. Towards evening the sound of the bagpipes announced the approach of the fine regiment, and Sir James was apprised of the fact. After some persuasion he consented to come forth, but no sooner was he seen by the men than they burst out into a cheer such as British soldiers only can give. Outram attempted to address them, but his sentences were interrupted by renewed outbursts, which so affected him that he could scarcely speak. An Italian officer, in the service of the Pasha of Baghdad, who was an eye-witness of this scene, remarked to an officer of the Force that he "should be sorry to command a whole division of Persians against one regiment of Highlanders."

The final march from Cawnpore to Lucknow, the terrible advance to the Residency, and the subsequent operations in which both General and men were engaged, had served to deepen and strengthen this friendly feeling. We have already seen that to the 78th had been allotted the post of honour both at the relief and the evacuation of the Residency; but Outram took an opportunity of still more emphatically expressing his high regard for the regiment. On the 26th of January 1858, while Outram's troops were posted at the Alumbagh, waiting for the arrival of Sir Colin Campbell's Siege Force, the second brigade was paraded to witness the presentation of six good-conduct medals to men of the 78th. Sir James addressed the regiment in terms of unqualified compliment, and later in the evening sent a letter to Brigadier Hamilton, who was their commander, from which we quote the principal passages :—

"Camp, Alumbagh, Jan. 26, 1858.

"My Dear Brigadier,—I should be sorry that the 78th should attribute anything I said to-day to the excitement of the moment, and therefore somewhat more perhaps than what I would deliberately record. What I did say is what *I really feel*, and what I am sure must be the sentiment of every Englishman who knows what the 78th has done during the past year ; and I had fully weighed what I should say before I went to parade.

While fresh in my mind, I will here record what I did say, in case you may think my deliberately and conscientiously expressed testimony to the merits of your noble regiment of any value.

The following is the spirit, and I think almost literally, what I said :—

"78th Highlanders—I gladly seize the opportunity of the brigade being assembled to witness the presentation of good

conduct medals to some of your deserving comrades, to say a
few words to you, to tell you of my high estimation of the
very admirable conduct of the whole regiment during the
year, completed to-morrow, that I have been associated with
you in the field, commencing in Persia.

" Your exemplary conduct, 78th, in every respect throughout the past eventful year, I can truly say, and *I do most
emphatically declare*, has never been surpassed by any troops
of any nation in any age, whether for indomitable valour in
the field or steady discipline in the camp, under an amount
of fighting, hardship, and privation, such as British troops
have seldom, if ever, heretofore been exposed to.

" The cheerfulness with which you have gone through all
this has excited my admiration as much as the undaunted
pluck with which you always close with the enemy whenever you can get at him, no matter what his odds against
you are, or what the advantage of his position ; and my
feelings are but those of your countrymen all over the
world, who are now watching your career with intense
interest.

" I trust it will not be long before the campaign will be
brought to a glorious conclusion, by the utter destruction of
the hosts of rebels in our front, on the capture of this doomed
city, their last refuge ; and I am sure that you, 78th, who
will have borne the brunt of the war so gloriously from first
to last, when you return to old England, will be hailed and
rewarded by your grateful and admiring countrymen as the
band of heroes, as which you so well deserve to be regarded."

After remarking that the " good and glorious" conduct of
the 78th had been nobly emulated by the other troops of
the division, especially by the 90th and 64th, Sir James
concluded by saying that he should never forget them.

Early in 1859, the regiment was ordered to return home. Before they left India, complimentary orders were addressed to them by Sir Robert Walpole and Sir Colin Campbell, both of whom spoke warmly in their praise. On their arrival at Bombay they found the place *en fête*. All Her Majesty's ships were covered with bunting, "rainbow fashion," and the Europeans ready to receive them with acclamations. In the evening the officers and men, with their wives and children, were entertained at a grand banquet. "A magnificent suite of tents," says Mr Keltie, "was pitched in the glacis of the fort, and many days had been spent in preparing illuminations, transparencies, and other decorations to add lustre to the scene. At half-past seven o'clock P.M. the regiment entered the triumphal arch which led to the tents, where the men were received with the utmost enthusiasm by their hosts, who, from the highest in rank to the lowest, had assembled to do them honour. After a magnificent and tasteful banquet, speeches followed, in which the men of the Ross-shire Buffs were addressed in a style sufficient to turn the heads of men of less solid calibre."

When they arrived at Fort-George, their headquarters, near Inverness, they were received with demonstrations of enthusiasm by their grateful and proud countrymen. At Brahan Castle they were feted by the Hon. Mrs Stewart Mackenzie, daughter of the Earl of Seaforth, who had raised the regiment. The town of Nairn paid the regiment the same compliment, and the noblemen and gentlemen of the northern counties did the men honour in Inverness. Subsequently they reached Edinburgh, where, if possible, they were received with still greater enthusiasm. They were hailed as "The Saviours of India," and banqueted in the Corn Exchange.

We may mention, in passing, that the Honorary Colonel of the Regiment, General Sir William Chalmers, K.C.B., died in June of that year at Dundee.   He was an able and distinguished soldier, who had seen much service, and fought many hard battles in the Peninsula.

In 1861 a monument to the memory of the officers, non-commissioned officers, and men of the 78th who had fallen

MONUMENT TO THE 78TH ON CASTLE ESPLANADE, EDINBURGH.

during the Mutiny, was erected on the Esplanade at the Castle of Edinburgh.   The monument is a striking and imposing erection, which never fails to attract the attention of visitors to the Castle.   It is in the form of a Runic cross, and bears an inscription and the names of those who fell. The inscription is in the following terms :—

" Sacred to the memory of the officers, non-commissioned officers, and private soldiers of the 78th Regiment who fell in the suppression of the Mutiny of the Native Army of India, in the years 1857 and 1858 ; this monument is erected as a tribute of respect by their surviving brother officers and comrades, and by many officers who formerly belonged to that regiment.—Anno Domini, 1861."

It is a pity the names and inscription had not been carved in plainer letters. The eccentric shape of the latter renders them almost illegible to many who would otherwise read them with interest.

CENTRE PIECE FOR OFFICERS' MESS OF 78TH, PRESENTED BY THE COUNTIES OF ROSS AND CROMARTY.

Shortly after the return home of the 78th, a meeting was held at Dingwall, the capital of Ross-shire, for the purpose of raising some testimonial to the regiment from the county which gave it its name. The result was the presentation of a magnificent centre piece, beautifully chased, for the officers' mess. It bears the inscription :—" Presented by the Counties of Ross and Cromarty to the 78th Highlanders, or Ross-shire Buffs, in admiration of the gallantry of the regiment, and of its uniform devotion to the service of the country— 1859."

New colours were presented to the regiment in 1868, and the old ones, which had seen so much service in the Mutiny, and in the protection of which so many gallant men had braved danger and yielded up life, were sent to Dingwall, where they were deposited in the Town Hall. There they remain, and, says Mr Andrew Ross, in his recently published " Old Scottish Colours," " there is not a stand in Scotland better or more vigilantly guarded. They are displayed on one side of the Town Hall, protected by glass from dirt or carelessness, the entire arrangement reflecting the highest credit on the municipal authorities."

The anecdotes preserved of the Mutiny are not numerous— the work was too stern to permit of much indulgence in humour. One, however, worth recording is told by James Grant in his " British Battles." During the advance by the 78th through the streets of Lucknow to the Residency a piper, who had been fighting his hardest through the *melee*, and paying little attention to words of command or to what was going on around him, suddenly discovered that he had lost his way in the smoke and dust, and that one of the enemy's cavalry was riding at him with uplifted sword to cut him down. He had fired his rifle, and had no time to fix his bayonet. A bright idea struck him. He seized his pipe, and putting it to his mouth blew forth such a wild, unearthly note that the fellow stopped as if shot, then turning his horse rode off at the gallop, leaving the piper to find his way safely back to his regiment.

We now leave the 78th, and go back to narrate the deeds of the other Highland regiments who took part in the work of suppressing the Mutiny. The first we are called upon to deal with is one with whose valour at the Alma and Balaclava our readers are already familiar—the gallant 93rd.

Chapter XIII.

THE INDIAN MUTINY—THE 93rd AT LUCKNOW.

THE 93rd, when it arrived in India to take part in the suppression of the Sepoy Mutiny, was scarcely, as General Shadwell describes it, "a regiment of seasoned veterans"—for it had received many recruits in its ranks since that glorious day on which it met and repulsed the charge of Russian horsemen at Balaclava—but it did possess its "full number of companies, and, judged by its Crimean antecedents, was of a quality not to be surpassed." It had returned from the Crimea sadly reduced by the stress and hazards of war, and in its ranks were many young eager men, new to warfare, but determined, at whatever cost, to maintain the glory of the grand old regiment. Among the recruits who had joined the headquarters of the 93rd at Dover, were over 400 officers and men who had formed the Dundee depôt. These had been a welcome accession of strength, and all the more welcome that they were a set of fine soldierly fellows, many of them drawn from Dundee itself; but a large number being true Highlanders, from the regions which lie to the northward of Juteopolis. They were commanded by Captain Middleton, and that their appearance had attracted attention is proved by a regimental order, dated 30th of August, in the following terms :—

"The officer commanding is desirous of expressing his high sense of the services rendered to the regiment by Captain Middleton, under whose command so fine a body of men has been trained at the depôt. Major-General Cameron, who inspected the depôt this day, was pleased to express his satisfaction with their appearance. To have deserved the good opinion of so experienced an officer must be considered most creditable to all ranks of the depôt."

Those youths, full of ardour, were, like the other troops in the field, impressed with the justice and honour of their cause, and determined that, in whatever conflict they might engage, they would be true to the traditions of the regiment, and to the promise of greatness to be achieved which their appearance had suggested to the Queen when parting with them in England. The veterans, with their medals on their breasts, needed no sentimental incentive to duty. Already in the Crimea they had won high honours in battle, and evoked the admiration of a wondering army. They were there to uphold the *prestige* they had gained. Their work but needed to be shown to them—they knew how it should be accomplished.

The regiment was under the command of Colonel Leith Hay, C.B., a gallant officer with whom the men had seen much service. As a subaltern in the 93rd he had gone through the Canadian Rebellion in 1838, and as Major he had been present at the battles of Alma and Balaclava, and had commanded the 93rd in their operations at the final assault on Sebastopol. He was a Scotsman—a cautious but brave Aberdonian—and a man to inspire the confidence of those under him.

Lieutenant-Colonel Adrian Hope, son of Sir John Hope, of Peninsular fame, and afterwards Earl of Hopetoun, Lieutenant-Colonel Ewart (now General Ewart, C.B.), and Lieutenant-Colonel Gordon (now Major-General Gordon, C.B.), were also with the regiment, and it may safely be said that braver men never led soldiers to battle. Adrian Hope was, indeed, a conspicuously able and dashing

soldier, and as a brigade leader earned much distinction
in the subsequent operations of the campaign in which he
lived to take a part. His brilliant career was, however,
all too early cut short. He fell in action at Rooyah on the
16th of April 1858, five months after his memorable exploit
at the Shah Nujjif.

BRIGADIER-GENERAL ADRIAN HOPE.

The regiment was hurried to the front in detachments,
and the first contact with the enemy took place at Kudjwa,
near Futtehpore, which will be remembered as the scene
of the first of the great series of victories achieved by
Havelock. On the 31st of October 1857, Lieutenant-
Colonel Gordon, commanding Nos. 3, 8, and Light Companies at Futtehpore, was suddenly summoned along with

his Adjutant, Ensign Dick-Cunyngham (now Sir R. Dick-Cunyngham, of Prestonfield), to meet Colonel Powell, the senior officer at Futtehpore, and Captain Peel, of the Naval Brigade, to consult as to the possibility of intercepting a large body of rebels—about 3,000 in number—who, while threatening the Grand Trunk Road between the British and Cawnpore, had the design of marching straight across country to Lucknow, there to assist the rebels who were harassing Outram and Havelock. The resolutions of the officers were quickly formed, and a flying column of 500 men, including 100 of the 93rd, was ordered forward with all haste. The column had twenty-four miles to cover, and marched at four o'clock on the following morning. At three in the afternoon the advanced guard, 50 men of the 93rd, had covered the distance, and found the enemy. Kudjwa was composed of many buildings and houses placed within walled enclosures, and in front was a high bank which almost amounted to a fortification. The advance guard immediately went forward in skirmishing order, and firing commenced. The skirmishers were quickly strengthened by the remaining company of the 93rd, who extended to the left of those already out. The skirmishers were now about 1,000 yards from the village, and the enemy could plainly be seen forming up for battle—their bayonets glittering brightly in the sun. As the 93rd, the Engineers, and the 53rd advanced, the main body of the enemy, preserving their formation, fired volley after volley with great steadiness, and also kept up a rapid fire of grape and round shot. By and by the fire of the British began to tell, the replies of the enemy became less steady, and their formation less perfect. At length the 93rd and 53rd got within striking distance ; they ceased to fire, and, gallantly led by Colonel Powell in person, levelling their bayonets, dashed at the guns. The old story was repeated. For a

moment the Sepoys gazed, first fascinated by the advancing
line of glittering steel, then terror-stricken at the terrible
aspect of the coming men.   They fired one or two ineffectual
shots, abandoned guns and position, and fled through the
village, the Highlanders pursuing and inflicting upon them
dreadful loss.   In two hours the battle was won, and the
victors were in possession of the camp, baggage, and guns
left by the rebels in their haste.   Many uniforms of murdered
English officers were found in the camp.   It was a victory,
but one which had cost dear—a fifth of the whole British
Force had been the sacrifice.   Colonel Powell had fallen shot
through the head, and three officers of the 53rd were down
with severe wounds.   Of the 93rd three only had been
killed, but Dick-Cunyngham and many men were wounded,
some dangerously, many severely.

Captain Burgoyne, the author of the recently published
" Records of the 93rd," relates an incident of the vicissitudes
of the wounded during this engagement.   Where they were
lying, under the charge of a few attendants, they were much
harassed by the fire from the bank on their left flank.
" While the main body of the Sepoys retreated through the
village, many of their stragglers crept round the flanks, and
the wounded were at one time in imminent danger from a
body of men under a leader of some note, who collected and
encouraged them to the attack.   This attack, however,
was averted at a most critical moment by the steadiness of
this handful of wounded men, who, maimed and bleeding
as they were, and none able to stand to wield a bayonet, at
the earnest command of a wounded officer who lay among
them, reserved their fire, until, just as the attack was about
to be made, a well-aimed volley crashed into them, killing
their leader (who was seen waving his tulwar on the top of
the bank as they came on) and some others, and effectually

prevented a repetition of the attack." This cool conduct of the wounded in a most critical position was referred to by the Commander-in-Chief in his remarks on the official report of the engagement. The charge of the Highlanders and the men of the 53rd had proved very disastrous to the enemy. Many of them, caught in their sin by the impetuous rush, were found dead near the guns, while the bodies of numerous others, overtaken in flight by their pursuers, strewed the streets of the village—the killed altogether amounting to fully three hundred. When the troops bivouacked at sundown, the 93rd had obtained their first experience of the actual hardships of the campaign, and now understood something of the nature of the sufferings under which the 78th had so bravely borne up. They had marched twenty-four miles unbreakfasted, they had fought a hard battle, and won a decisive victory; they had gathered in their wounded, and laid out their dead —and all they had had to sustain their strength was a poor ration of biscuit and rum. The dead they carried with them back to Futtehpore, and there they were sadly buried.

By the 7th of November the whole of the regiment was collected in the plain of the Alumbagh, where it was attached to the Fourth Brigade, commanded by their old Lieutenant-Colonel, Brigadier Adrian Hope. The other regiments of the Brigade were the 4th Punjab Rifles and the 53rd (Shropshire) Regiment. The whole army numbered about 4,200 men, and with it Sir Colin Campbell intended to relieve the beleaguered troops who still remained within the Residency at Lucknow—and to carry off the garrison, the wounded, women, and children.

On the 9th Sir Colin himself arrived in the camp, and immediately concerted measures for the work before him. The grim old soldier was once more about to rejoice in the stir and rage of battle; his love of warfare having suffered no diminution since his Crimean exploits. To most of the men of the Relieving Force he was personally a stranger; and wishing to form an opinion of their qualities, he held on the 11th a general review. Sir Colin addressed each corps in turn, and while he was speaking words of hope and encouragement, the men took the opportunity of studying and measuring their leader; and they drew from the observation that feeling of confidence and security in his leadership which his presence and fame never failed to inspire. On the extreme left of the little army was the Fourth Brigade, and in its centre, "in tall and serried ranks," stood the 93rd— "conspicuous," says General Shadwell, "for their magnificent appearance, both as regards numbers and physique." When Sir Colin approached the Highlanders he read the look of recognition and welcome in many a kindling eye, and saw the pleasure with which they hailed this union with him once more on the battlefield. They were of one nation and blood, and "the recollection of dangers confronted and hardships endured" together was fresh in the memory of the General and of every bronze-faced veteran present. This was almost immediately made manifest. Turning to the Highlanders, Sir Colin said—

"Ninety-third! we are about to advance to relieve our countrymen and countrywomen besieged in the Residency of Lucknow by the rebel army. It will be a duty of danger and difficulty, but I rely on you!"

He was answered with a burst of enthusiastic cheering. "A waving sea of plumes and tartans they looked, as, with loud and rapturous cheers, which rolled over the field, they welcomed their veteran commander, the chief of their choice." This expression of confidence and enthusiasm on the part of the Highlanders had an almost electric effect upon the other troops present, and as the veteran General

L

rode back to camp his progress was marked by cheer on cheer, raised in succession by each corps he passed. This pleasing incident occurred, as we have said, on the 11th of November, and, curiously enough, two days before, in far-off England, Queen Victoria had been writing to Lord Canning, and used these words :—" We are glad to hear such good accounts of Sir Colin Campbell. . . . We can well imagine his delight at seeing his gallant and splendid 93rd, whom we saw at Gosport in June, just before they left."

On the 14th all was ready for the attack, and at six o'clock in the morning the troops stood to their arms. The task before them was known to be a severe one. The Sepoys numbered at least 50,000, and had strongly fortified themselves within the city. The Dilkoosha and Martiniere College, two large buildings with spacious grounds, were held in strong force as rebel outposts, and these Sir Colin determined at the outset to possess. He was then to attack the Secundra-Bagh, the Shah Nujjif, and other fortified buildings lying between him and the Residency; and, capturing these, secure a safe path of retreat for the beleaguered garrison.

It was nine o'clock before the troops were on the move. The 93rd was in the main column, with the exception of 200 attached to the rearguard, which was under the command of Lieutenant-Colonel Ewart, of the 93rd. As the head of the column neared the Dilkoosha, a heavy musketry fire was opened on it from the left, but this was quickly silenced. Then a large body of the enemy advanced through a wood inside the Dilkoosha Park, and opened fire; they, too, however, rapidly fell back, with the head of the column in pursuit. But now the British were within sweep of the fire from the enemy's outlying positions, and the roar of a general engagement soon filled the air. The British field guns and heavy batteries were playing, and the skirmishers were advancing, cheering and firing as they went.   In a short time they had driven the enemy beyond the line of the Canal, and the 93rd, piling arms, enjoyed a brief period of rest, under cover of some old mud walls to the right of the Dilkoosha.   Then the order was given to attack the Martiniere.   In turn it was captured, the regiment, greatly divided, being hotly engaged the whole time of the conflict.

When night fell it found the little British Force in possession of important positions, and there the troops lay down in the open air for the night.   The duties of the regiment on the day following, the 15th, if lacking in dash and brilliance, made large demands on the men's steadiness and courage.   The headquarters position—for the most part extended along the bank in rear of the Martiniere compound opposite the Canal—was exposed to an annoying and incessant, if not very destructive, musketry fire.   Yet this sort of work, familiar enough to those who had done duty in the Crimean trenches, served to harden the younger soldiers, and fit them for the work that lay before them.

And on the morrow their fighting qualities were put to the test.   The Dilkoosha and the Martiniere were now secure, but beyond lay the Secundra-Bagh and the Shah Nujjif.   Strongly built and fortified, and held by immense numbers of rebels, both positions appeared almost impregnable ; and all knew that their capture must involve a terrible sacrifice of life.   During the night of the 15th, the 93rd had bivouacked close under the Martiniere, and at six in the morning of the eventful 16th, they, along with the rest of the troops, were under arms.   Before nine they were advancing—this time the Highlanders' occupying the proud position of leading regiment in the main column proceeding to storm the Secundra-Bagh.   First they moved resistlessly along by the bank of the Goomtee, driving back the enemy; then, making a sharp turn to the left, they met the full force of the enemy's resistance. From enclosures and huts, and from the great building filled with thousands of rebels, was poured into their ranks a raking and deadly fire. For an hour the relative positions of the opposing forces were but little changed, although the hot conflict raged without intermission. To Infantry the Secundra-Bagh, held by such numbers, was perfectly inaccessible; and they could go no further forward till Blunt and Travers' guns pounding at the building had made a breach in its walls.

With eager impatience the Highlanders watched shot after shot flatten itself, and rebound without doing substantial damage to the structure; while the enemy's bullets directed in showers against themselves only too often found their mark. At length a piece of the masonry near the base of one of the corner towers was seen to be yielding, and at the weak point another and yet another shot was fired, till, amidst breathless excitement, it finally gave way, and a breach was visible. Now was the time for action. The bugles sounded the assault, and the 93rd, who during the cannonade had been seeking the best shelter they could find, sprang to their feet.

Chapter XIV.

JUST before the bugle sounded the advance, a young Scotch officer—endowed with all the ambition and thirst for glory belonging to his class, and possessing rather more than the ordinary share of daring and athletic vigour—had been standing at a gap in a mud wall, behind which the 93rd were lying. He had not been idle in the period of inaction during which the big guns pounded at the walls of the Secundra-Bagh. Accustomed to the use of the gun and rifle on his Orkney moors and pastures, he had, with weapons loaded and handed to him by his men, been keeping up his practice by firing at the copper-coloured faces showing through the loopholes of the building. This was Captain F. W. Traill-Burroughs, of No. 6 Company, who had already seen nearly ten years service with the regiment, and who had at the Alma and Balaclava, and in all the subsequent services of the Crimean Campaign, proved himself a cool, courageous, and promising officer. The opening of the breach, and the sudden order to the 93rd to advance, made him at once give up his rifle practice, and, flinging from him the smoking weapon, he drew his sword, and shouting "Forward, No. 6," dashed right across the open ground to the breach, which was almost straight in his front. Wildly as the Highlanders and Sikhs had rushed forward, shoulder to shoulder in the race, Burroughs outdistanced the whole, and was the first to arrive at the breach. He could not get through. The hole was too small to permit of a man's passage ; but to clear away some of the now loose and crumbling masonry was the work of but a second or two, and in a minute he had scrambled through and tumbled headlong into a room full of excited and now startled Sepoys. His feather bonnet fell from his head, and rolled towards the mutineers, who, probably taking it for some fearful instrument of war, fired one or two ineffectual shots at the daring intruder, then hurried from the room. By the time Burroughs had picked up and adjusted his headgear, three men—Corporal Robert Fraser, Lance-Corporal John Dunlay, and Private William Nairn—had got through the breach behind him. Followed by these, the daring young officer dashed after the enemy in the direction of the gateway marked as B in the accompanying plan, drawn from the one furnished by General Burroughs for Burgoyne's "Records of the 93rd." When they had gone about half-way they were assailed by a large number of the enemy, before whom, as Dunlay had already been wounded, they slowly retired to A, firing as they went. By this time more men of No. 6 Company were through the breach ; and, led forward once more by Burroughs, they advanced to the gates with the intention of opening them to the troops outside. Here a desperate struggle ensued—the fighting was hand-to-hand—a few brave British soldiers amidst hundreds of Sepoys. As some of the men worked at the gates, the others kept up the unequal combat, Burroughs being among the latter. In the struggle, however, he was unfortunately cut down by a mutineer who attacked him while he was engaged with another. He received a slashing blow from a tulwar, which cut open his right ear and cheek —his life being saved only by the resistance offered by his feathered bonnet, which was dented "like a bishop's mitre" by the stroke. As the brave fellow sank under his wound his men succeeded in forcing open the gates, and in poured officers and men of the 93rd, 53rd, and Sikhs, who had been clamouring without—Sir Colin himself among the number. Seeing Burroughs faint and covered with blood, Sir Colin asked if he was much hurt, and being assured by the gallant officer that he was not, he passed on to superintend the attack.

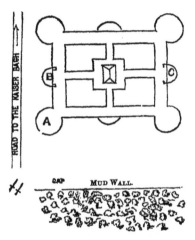

GROUND PLAN OF SECUNDRA-BAGH.

Having described the leading part taken by Captain Burroughs in the storming of the Secundra-Bagh, we now return to notice the heroic conduct of other officers and men of the 93rd.

When the regiment sprang forward, obedient to the command of Colonel Ewart, Ensigns Robertson and Taylor gave the colours of the regiment to the breeze, and, bounding over the mud wall, endeavoured to head the advancing line. But others crowded before them; in the wild and exciting rush the strong-limbed mountaineers doing their utmost to outstrip the fleet-footed and nimble Sikhs. But outrunning others in this general advance was Lieutenant Richard Cooper, of No. 5 Company—a strong athletic youth—who, reaching the breach already opened out by Burroughs, bounded through it with one great leap, and landed inside unscathed. Following on his heels came Colonel Ewart, Captain Lumsden, and eleven Highlanders and Sikhs; but the others, impatient at the delay which must ensue in going one by one through the small aperture, abandoned the attempt to enter by the breach, and went round to force the gate, which Burroughs and his followers were by this time, from the inside, struggling to open—if, indeed, they had not already opened it.

Ewart and Cooper found themselves in an enclosure of "one hundred and fifty yards square, with towers at the angles, and in the centre of the eastern face a building, consisting of a room opening out of a courtyard behind it." The grass growing all over the ground of the enclosure was sufficiently high to conceal the enemy from view. Instead of taking the path to the left, along which Burroughs had proceeded, Ewart, Cooper, and the others dashed along that leading to the right towards C in the sketch. They were in presence of large numbers of the enemy, who, from many points of vantage, were firing upon them. Lumsden, a brave Aberdonian, had already been shot dead as he was cheering on his countrymen, waving his sword above his head, and calling—"Come on, men, for the honour of Scotland!" When the others had reached the front of the square building already referred to, they found the enemy in still greater numbers. "There were," says Colonel Malleson, "rebels in front of it, rebels within it, rebels in the court-yard behind it. But on this occasion," continues the Colonel, "boldness was prudence. The rebels outside, astonished by the sudden appearance of the two British officers and their following, ignorant of their numbers, and believing, it may be presumed, that the main entrance had been forced, ran hurriedly into the building, and attempted to make their way through a small door into the courtyard behind. The two officers and their men dashed after them, and a hand-to-hand encounter ensued."

Cooper went straight into their midst. In his own words, he "worked away" at them with his whole strength, until he had laid many low with his sword. Then he was singled out by a Native officer—one of the Loodiana regiment which had mutinied at Benares—bearing a shield in his left hand and a tulwar in his right. Dropping his shield, the rebel—a very tall man—cut straight at Cooper, who, however, at the same moment had aimed a blow at his antagonist. The tulwar cut through Cooper's feather bonnet and deep into his head, but the Highlander's sword had struck home, and the rebel fell dead. Cooper, too, sank upon the ground. With him the struggle was over, unconsciousness supervened, and he knew no more till he had been carried beyond the region of the strife. But he had still the pluck left to read with mended head the burial service of a brother officer— Captain Jas. Dalzell—the same evening. When Cooper fell, Ewart was engaged in a desperate encounter with a number of rebels. Followed by his men, he had entered the courtyard and dashed at the rebels congregated there. As the Highlanders approached, the rebels delivered a volley at ten paces distance. Fortunately, they fired too high, missing all but Ewart's bonnet, which was pierced by a ball. The fighting became desperate. The men bayoneted all who opposed them, but one rebel was no sooner down than others were crowding on to fill his place. A powerful Sepoy was making

at Ewart, when that officer shot him dead with his revolver.
With the same weapon he killed five others who followed
the first. Then his sword was wielded with a will. " Still,"
says Malleson, "numbers might have prevailed, when, at the
critical moment, the Highlanders, the Sikhs, and the 53rd
pressed in to the rescue"—entering by a second breach which
had been made in a window on the right, in the vicinity of C.

The rebels were now driven back on all sides; but from
the towers at the angles of the enclosure they opened a terrific
musketry fire, almost at point blank range. This they diversi-
fied by occasionally descending and engaging in close hand-
to-hand combat; their numbers enabling them to inflict
many casualties in the British ranks. While this conflict was
going on, Colonel Ewart once more distinguished himself by
a daring feat. Observing a colour in one of the rooms into
which the rebels had retreated, he determined to possess it,
and instead of directing other men to perform the hazardous
task, he resolved to do it himself. Quite unassisted, he
made a dash at the room, to find the entrance defended
by two Native officers. With these he at once engaged.
The conflict was stubborn. Twice Ewart was wounded, but
he clung to his purpose, and fought on till both men lay
dead at his feet. Then seizing the colour, he carried it
proudly forth, and the same evening presented it to Sir Colin
Campbell as a trophy of victory.

It was rather before this juncture that a brilliant feat of
arms was accomplished by one of the most popular officers
of the regiment—Captain W. G. Drummond Stewart, of
Murthly. Stewart was of a free-and-easy disposition,
generous to a fault, brave as a lion, and with the reputation
of being remarkably self-possessed in action, nothing in the
slightest degree disturbing his equanimity. Two guns were
raking the road in the vicinity of where he was stationed

with Nos. 2 and 3 Companies of the 93rd, whose object was to keep down the flank fire of the enemy. The guns proving decidedly troublesome, Stewart proposed to capture them, and, supported by a small party of the 93rd and 53rd, who had readily volunteered to follow him, he dashed forward with great spirit and daring. After a short struggle the attack proved entirely successful, and had the important result of securing the Mess-house, and enabling connections to be formed between this post and the Secundra-Bagh. For this gallant deed Stewart was elected by the officers of the regiment to receive the Victoria Cross.

Meanwhile the fighting fiercely raged within the enclosure. The rebels, pressed from point to point, turned at bay and fought with the desperation of despair. Every staircase, corridor, and room was contested. The British were filled with a vengeful desire to bring retribution upon the miscreants for their long record of bloody misdeeds, and the word "Cawnpore" was hissed into many a dying wretch's ear. Never had the avengers such opportunity as this. There was no chance of flight for the rebels here. This time they would have to fight to the bitter end. Quarter was neither asked nor given. That building was to be the huge sarcophagus of those on the one side or the other. Nor was there long any doubt as to which side would triumph. The power of the British arms was irresistible. Each man fought with a courage and fury which sometimes struck wonder into the hearts of his opponents, and even paralysed their efforts at resistance. For hours the battle went on, nor did it end while one single rebel stood to level a musket or brandish a tulwar. When at last, at three o'clock in the afternoon, the British stood conquerors in the Secundra-Bagh, around them were a sickening sea of blood and piles of dead—over two thousand Sepoys having fallen and died where they fought.

Of the whole number who garrisoned the Secundra-Bagh at the commencement of the engagement, it was said that only four escaped, and Colonel Malleson thinks that the escape of even these four is doubtful.

From such a desperate fight it could not be expected that the Highlanders would escape scatheless. Burroughs, Cooper, and Ewart, as we have already seen, were wounded early in the action. Lumsden was killed; Captain Dalzell shared his fate. Lieutenant Welch was severely wounded by a musket shot, and Ensign Macnamara received a sword-cut on the head. Donald Murray, the sergeant-major of the regiment, shot dead, had been one of the first to fall; and many of the rank and file were killed and wounded.

Of the rewards given to the regiment for their heroic conduct on this occasion, we have already mentioned the Victoria Cross bestowed upon Captain Stewart. Another Victoria Cross was given to Lance-Corporal J. Dunlay, " for being the first man now surviving of the regiment who, on the 16th of November 1857, entered one of the trenches of the Secundra-Bagh, at Lucknow, with Captain Burroughs, whom he most gallantly supported against superior numbers of the enemy." The names of Colonel Ewart and Captain Cooper were voted upon by the officers of the regiment for recommendation for the Victoria Cross along with that of Captain Stewart, but Stewart received the majority of votes, and obtained the distinction. Victoria Crosses were likewise gained by Privates David Mackay and Peter Grant, and Sergeant Munro received the same honour for distinguished conduct within the enclosure.

Captain Burroughs was also recommended by Colonel Leith Hay for the Victoria Cross—" for individual gallantry in the Secundra-Bagh, being *the first* who entered one of the breaches, and engaged in personal combat with greatly

superior numbers of the enemy, in which he was wounded
by a sword-cut." The recommendation was laid before Sir
Colin Campbell, but we regret to say did not receive that
attention which it certainly merited. It was supported by
the Brigadier, Adrian Hope; but Sir Colin, like many
smaller men, had his prejudices. In the first place—
apparently forgetting a certain incident of his youth at St
Sebastian—he had begun to dislike the rush and dash made
by young men for the Victoria Cross, and was inclined to
discourage special efforts which he thought were made solely
to obtain the coveted honour. In the next place, Sir Colin
stuck stubbornly to his own judgments, and Captain
Burroughs had been unfortunate enough in his wounded
state to meet and speak with the Commander-in-Chief as the
latter entered the Secundra-Bagh by the gateway, which
Burroughs and his followers had succeeded in clearing from
the inside. When, therefore, he read that the Captain was
said to have entered by the breach, he declared it impossible,
maintaining that he had seen him at the gateway, and that
he must have entered it with the others who went that way.
It was useless for Hope to attempt to explain the matter.
Sir Colin would listen to nothing, and was so entirely
satisfied with the conclusions derived from what he considered
his own knowledge that he did not forward the recommendation, and so destroyed the Captain's chance of obtaining the
prize dearest to the soldier's heart.

However galling and disappointing it may be to a brave
officer to be denied the honour he has worthily won, it is
still worse to be refused the credit of having accomplished the
feat for which the honour is sought. Yet General Burroughs
has undoubtedly suffered in this respect. Colonel Malleson,
while rather ostentatiously going out of his way to obtain
historical justice for Ewart and Cooper, and giving what

purports to be a full description of the conduct of those who
first entered the Secundra-Bagh, strangely enough ignores the

Lieutenant-General F. W. TRAILL-BURROUGHS, C.B., of Rousay, Orkney.

very name of Burroughs.   He writes as if he were unaware
of the existence of any claim of General Burroughs to the

distinction. As he also ignores the name of poor Lumsden, by whom General Ewart has stated he was accompanied, the credibility of his narrative is considerably impaired. It suffers further from the declaration of Ewart himself that he saw Burroughs already in the building when he jumped through the hole. As Ewart went in with Cooper, and was followed by Lumsden, two facts become apparent—(1st) That General Burroughs was beyond question before these officers in the Secundra-Bagh ; and (2nd) that in ignoring his claim to the honour, and writing up the claim of another—however brave and distinguished that other may be, or however much he may have suffered from official neglect—Colonel Malleson not only does not take the highest view of historical justice, but does an able and gallant soldier a positive wrong. The fact of Burroughs being at the gap of the mud wall, and not having to climb over it like the others, gave him the start in the race, and his going to the left and Cooper to the right when the Secundra-Bagh was entered is probably the explanation of their not meeting, and the confusion of claims arising. We may just add that since Corporal Dunlay was considered worthy of the Victoria Cross for being the first to follow Captain Burroughs, it is obvious that Sir Colin Campbell's stubbornness in refusing to transmit Colonel Leith Hay's recommendation could not be justified upon the ground that Burroughs, having led where the other followed, had not a paramount claim to the distinction.*

Having digressed thus far to substantiate our own account of the storming of the breach, we now turn to describe the further work of the 93rd on the eventful day of which we have been writing. Severe as had been the struggle in the Secundra-Bagh, a still more terrible combat had immediately to be undertaken by the regiment.

* See Appendix E.

Chapter XV.

THE event which followed the taking of the Secundra-Bagh is brilliantly described by the author of " Lord Clyde's Campaign in India," an article which appeared in *Blackwood's Magazine* for October 1858. The writer, who was present during the operations, is said, upon good authority, to have been Sir Archibald Alison, who was at that time Military Secretary to the Commander-in-Chief.* His description has been so much drawn from by later writers, and so mangled and changed, that we believe we do fuller justice, both to him and ourselves, by simply placing it within inverted commas, and quoting it almost entire.

" After passing between the Seria and the Secundra-Bagh," he says, " the road to the Residency leads straight across an open plain about twelve hundred yards broad. About three hundred yards along the road there is a small village, with garden enclosures round it, while about two hundred and fifty yards further on, and one hundred yards to the right of the road, stood the Shah Nujjif, a large mosque, situated in a garden enclosed by a high loopholed wall. This wall is nearly square, and very strong. Between it and the plain is a thick fringe of jungles and enclosures, with trees and scattered mud cottages, which make it impossible to get a distinct view of the place till you are close on it. Between it and the Secundra-Bagh, amidst jungles and enclosures to our right of the little plain, was a building on a high mound, called the Kuddum Russul.

* See Appendix F.

"Hope having now drawn off his brigade from the Secundra-Bagh, led it against the village, which he cleared and occupied without much difficulty ; while Peel brought his 24-pounders, mortars, and rocket frames, and placed them in battery against the Shah Nujjif in an oblique line, with their left resting on the village. The musketry fire which streamed unceasingly from that building was most biting and severe ; and after nearly three hours' battering, it was still unsubdued. An attempt, made with great gallantry by Major Barnston, with his battalion of detachments, to drive the enemy from the fringe of the jungle and enclosures in front, by setting fire to the houses, proved unsuccessful ; but on the right the Kuddum Russul was assaulted and carried by a party of Sikhs.

"In the narrow lane leading up from the rear meanwhile the utmost confusion prevailed. The animals carrying the ordnance and engineer supplies, unable to advance from the enemy's fire in front—unable to get out on either side, and pressed forward by those in rear—got completely jammed, insomuch that an officer, sent to bring up ammunition and all Greathed's disposable Infantry to the now hard-pressed front, had the utmost difficulty to get the men along in single file ; whilst, some houses having been wantonly set on fire by the camp followers, the passage was for a time entirely blocked up ; and it was only when the flames were abating that a string of camels, laden with small-arm ammunition, which was urgently required by the troops engaged, could, with great risk and toil, be forced through the narrow and scorching pass. Even then, however, the confusion near the Secundra-Bagh had got to such a pitch that all passage had become impossible ; and had it not been that a staff officer discovered a by-path leading into a broad road, which abutted on the Secundra-Bagh, neither men nor ammunition could have been brought up.

M

"In front of the Shah Nujjif the battle made no way. The enemy, about four o'clock, got a heavy gun to bear upon us from the opposite bank of the river, and their very first shot blew up one of Peel's tumbrils, whilst their deadly musketry had obliged him to withdraw the men from one of his pieces, and diminished the fire of the others. The men were falling fast. Even Peel's usually bright face became grave and anxious. Sir Colin sat on his white horse, exposed to the whole storm of shot, looking intently on the. Shah Nujjif, which was wreathed in columns of smoke from the burning buildings in its front, but sparkled all over with the bright flash of small arms. It was soon apparent that the crisis of the battle had come. Our heavy Artillery could not subdue the fire of the Shah Nujjif—we could not even hold permanently our advanced position under it. But retreat to us there was none; by that fatal lane our refluent Force could never be withdrawn. Outram and Havelock and Inglis, with our women and children, were in the front, and England's honour was pledged to bring them scatheless out of the fiery furnace. What shot and shell could not do the bayonets of the Infantry must accomplish. But the crisis was terrible. Even as the fate of the French Empire hung at Wagram on the footsteps of Macdonald's column, so did the fate of our Indian dominions depend that day on the result of the desperate assault now about to be undertaken.

"Collecting the 93rd about him, the Commander-in-Chief addressed a few words to them. Not concealing the extent of the danger, he told them that he had not intended that day to have employed them again, but that the Shah Nujjif *must be taken;* that the Artillery could not bring its fire under, so they must win it with the bayonet. Giving them a few plain instructions, he told them he would go on with them himself.

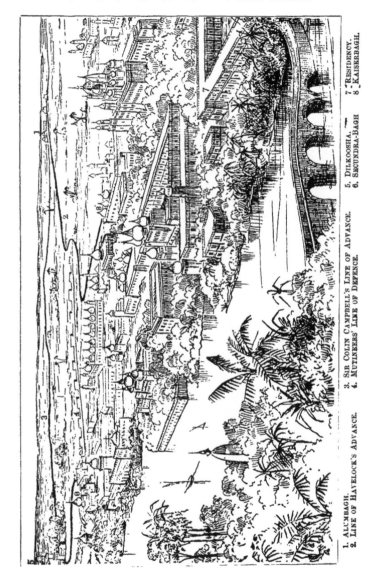

1. ALUMBAGH.
2. LINE OF HAVELOCK'S ADVANCE.
3. SIR COLIN CAMPBELL'S LINE OF ADVANCE.
4. MUTINEERS' LINE OF DEFENCE.
5. DILKOOSHA.
6. SECUNDRA-BAGH.
7. RESIDENCY.
8. KAISERBAGH.

" To execute this design, Middleton's battery of the Royal
Artillery was ordered to pass Peel's guns on the right, and,
getting as close as possible to the Shah Nujjif, to open a fresh
and well-sustained fire of grape. Peel was to redouble his
fire, and the 93rd to form in column in the open plain close
to the village ready to rush on.

" Middleton's battery came up magnificently with loud
cheers, the drivers waving their whips, the gunners their
caps. They galloped forward through the deadly fire to
within pistol shot of the wall, unlimbered, and poured in
round after round of grape. Peel manning all his guns,
worked his pieces with redoubled energy, and under cover of
his iron storm the 93rd, excited to the highest degree, with
flashing eyes and nervous tread, rolled on in one vast wave.
The grey-haired veteran of many fights rode with his sword
drawn at their head; keen was his eye as when in the pride
of youth he led the stormers at St Sebastian. His staff
crowded round him. Hope, too, with his towering form and
gentle smile, was there, leading, as ever was his wont, the
men by whom he was loved so well. As they approached
the nearest angle of the enclosure the soldiers began to drop
fast, but without a check they reached its foot. There, however, they were brought to a stand. The wall, perfectly
entire, was nearly 20 feet high, and well loopholed; there
was no breach, and there were no scaling ladders. Unable to
advance, unwilling to retire, they halted and commenced a
musketry battle with the garrison, but all the advantage was
with the latter, who shot with security from behind their
loops, and the Highlanders went down fast before them. At
this time nearly all the mounted officers were either wounded
or dismounted. Hope and his aide-de-camp were both
rolling on the ground at the same moment with their horses
shot under them; his Major of Brigade had just met with

the same fate ; two of Sir Colin's staff had been stricken to the earth ; a party who had been pushed on round the angle to the gate had found it so well covered by a new work in masonry as to be perfectly unassailable. Two of Peel's guns were now brought up to within a few yards of the wall. Covered by the fusilade of the Infantry, the sailors shot fast and strong, but though the masonry fell off in flakes, it came down so as to leave the mass behind perpendicular and as inaccessible as ever.

"Success seemed now impossible. Even Hope and Peel, those two men iron of will and ready of resource, could see no way. Anxious and careworn grew Sir Colin's brow. The dead and wounded were ordered to be collected and carried to the rear. Some rocket frames were brought up, and threw in a volley of these fiery projectiles with such admirable precision that, just skimming over the top of the ramparts, they plunged hissing into the interior of the building, and searched it out with a destroying force. Under cover of this the guns were drawn off. The shades of evening were falling fast—the assault could not much longer be continued. Then, as a last resource—the last throw of a desperate game—Adrian Hope, collecting some fifty men, stole silently and cautiously through the jungle and brushwood away to the right, to a portion of the wall* on which he had,

* The spot in the wall was really discovered by Sergeant John Paton, of the 93rd. Paton then guided Hope and his party to the spot, through which an entrance was effected. The gallant sergeant was rewarded with a Victoria Cross for this important and serviceable deed. John's father was a soldier, and he was born in Malta. He joined the 93rd Highlanders at an early age, and was present with the regiment at the Alma, Balaclava, and the other engagements taken part in by the regiment during the Crimean War. He subsequently went through the many battles of the Mutiny in India in which the 93rd were engaged, including the battle of Cawnpore, the affair of the Kalee Nudee, the final storming of Lucknow, the battle of Bareilly, &c. He was a quiet, intelligent, sober-living, as well as brave and hardy soldier. After he left the regiment, to which his services had proved of so much value at the time of pressing need, he received the appointment of chief warder in the gaol at Port M'Queen, New South Wales. That position, we believe, he still occupies. The portrait we give next page is from a photograph taken some years ago. It shows him with his medals and V.C. on his breast.

before the assault, thought he perceived some injury to have
been inflicted. Reaching it unperceived, a narrow fissure
was found. Up this a single man was, with some difficulty,
pushed. He saw no one near the spot, and helped up Hope,
Ogilvy (attached to the Madras Sappers), Allgood, the
Assistant Quartermaster-General, and some others. The

SERGEANT JOHN PATON, V.C.

numbers inside soon increased, and as they did so they advanced, gradually extending their front. A body of Sappers,
sent for in haste, arrived at the double; the opening was
enlarged, the supports rushed in. Meanwhile Hope's small
party pushing in, to their great astonishment, found themselves almost unopposed. Gaining the gate they threw it
open for their comrades. The white dresses of the last of the

garrison were just seen gliding away amidst the rolling smoke into the dark shadows of the night. Panic-struck, apparently, by the destruction caused by the rockets, and the sudden appearance of some of the assailants within the walls, they fled from the place, and gave up the struggle just when the victory was secure.

" Never had there been a harder fought day, but never was a result gained more satisfactory. The relief of the Residency, an affair in the highest degree problematical in the morning, was now all but certain. To-day we had fought for existence, to-morrow we would throw for victory. Anxiously during the day had we strained our eyes, frequently had we strained our ears, looking to gain some indication of Outram's progress, looking wearily for the flash of his musketry in the enemy's rear, to lighten us from the load under which we were staggering. But hour after hour passed by, and no bickering tide of war came up to us from the east. But there was now joy in every heart—there was light in every eye. Not in vain now had Greathed's toil-worn bands pushed on in hot haste from Delhi's smoking ramparts—not in vain had the Highlanders hastened over the stormy main from their distant mountain homes—for the blood of our defenceless women would not now ascend reeking to the heavens."

The day's fighting ended, the tired soldiers sank down to rest in their ranks. They had performed a work which, in his despatch, the Commander-in-Chief described as an "action almost unexampled in war." Their sleep was fitful and broken. All round were the enemy bent on harassing the now much-extended British line. In the early morning the regimental flag of the 93rd was placed on the highest pinnacle of the Shah Nujjif, and given to the breeze as a signal to the garrison beyond that the place had been reduced, and that

Sir Colin was surely, if slowly, making progress. The task
of placing the flag was performed by Lieutenant M'Bean, the
Adjutant of the regiment, one of the bravest of the brave,
and one of whom we shall yet have to record brilliant
things. M'Bean was assisted by another Highlander, Sergeant
Hutcheson. When the enemy saw their purpose they opened
a heavy fire on the two men ; but the heroes calmly finished
their work amidst the whizzing and whistling of the bullets,
and retired uninjured.

In the fighting, renewed with dauntless spirit on the 17th,
the 93rd had no more difficult duty than to hold the posts
they had already with such conspicuous gallantry captured.
The other troops pushed the relieving movement to success.
The Mess House* was taken, the Motee Mahul was shortly
afterwards in their possession ; the communication with the
Residency was open. Yet the enemy was strong, active, and
vigorous ; he was posted within musket shot of the Residency
gate ; but, anxious to welcome those who had come to their
deliverance, Outram and Havelock rode out, as they had
ridden in, through the storm of lead, and met the grey-haired
Commander-in-Chief who had hazarded so much to effect
their relief. When the Generals had met, the fact was
signalled to the British Force, and from none did heartier
cheers arise than from the lips of the 93rd. For the next
few days there was much desultory firing ; by the night of
the 22nd the rebels were so far held in check that it was
deemed safe to remove the garrison and abandon the Residency.

* It is worthy of note that the Mess House was taken by a storming party of the
90th Regiment, under the command of Captain (now Lord) Wolseley, and, in the
words of Colonel Malleson, " never was a daring feat of arms better performed."
He drove out the enemy by much hard fighting, and then exceeded his instructions by following them in their flight, instead of holding the post he had gained.
But by following the rebels he captured the Motee Mahul, and thus modified the
resentment of Sir Colin, who severely reprimanded him for his conduct. To his
reprimand, however, Sir Colin added a compliment to the young officer's courage
and ability, and congratulations on his success, and finally promised to recommend him for promotion.

By the morning of the 23rd, every man, woman, and child had been safely removed—through the streets of a hostile town, with an enemy on all sides numbering ten to one of the Relieving Force—every gun that was of use was carried away, with all the treasure and stores the Residency contained ; and everything was at length safe within the British camp at the Dilkoosha. Thus was at last accomplished that for which Havelock and Outram and the gallant Highlanders had so nobly fought, and which had cost the loss of so many valuable lives. Yet, as we mentioned in our remarks on the 78th, Lucknow was not yet subdued, and Sir Colin, leaving Outram with 4,000 men well posted in charge of it, went off to fresh conquests, to return in many days and strike the final crushing blow at this hotbed of rebellion.

During the hardest of the fighting at the Shah Nujjif, Colonel Leith Hay had his horse disabled by a musket shot, Hope had his killed under him, and at the same moment his aide-de-camp (young Lieutenant Butter of Faskally) tumbled to the ground from the same cause. Major Alison, the writer from whom in this chapter we have largely quoted, was so severely wounded in the arm that the limb had to be amputated. When the Artillery was brought forward to almost the base of the enclosure, the men of the 93rd aided the gunners in dragging the guns. In this the officers readily lent a hand—Hope, Colonel Hay, and Sir David Baird setting a splendid example to the men.

The losses of the 93rd from the 12th of November to the 22nd were heavy—including over 30 killed, and between 70 and 80 wounded and missing. "Two very unfortunate events," says Captain Burgoyne, "occurred on the evening of the 23rd. A corporal and three men were terribly burnt by the accidental explosion of a quantity of gunpowder which had been left on the ground, and all died in a day or two

EAST VIEW OF RESIDENCY AT LUCKNOW.—AFTER THE EVACUATION.—*From a Photograph.*

afterwards. The accident is said to have been caused through a spark falling from a soldier's pipe. And Colour-Sergeant David Knox was lost. He had formerly been in the 78th Highlanders, which regiment having come out of the Residency, was close by. He had been mustered in the morning, and naturally enough went to see his old comrades. He continued absent so long that inquiries were made for him, when it was ascertained that he had remained some time with the 78th, and left them just before dawn, saying he must return to the 93rd. From that moment nothing was ever heard of Knox. He may have missed his way and fallen into the hands of the enemy ; but it is supposed to be more probable that in the uncertain light he fell into one of the many deep wells which abound at Lucknow.

So silently and secretly had the withdrawal of those in the Residency been effected, that the enemy were not aware of our posts being deserted till many hours after. Alexander Macpherson, one of the sergeants of the 93rd, was accidentally left behind in the barracks. After the roll had been called, before two o'clock on the morning when the British army and its precious charge retired, Macpherson, tired and worn out, fell asleep. On waking, he was surprised to find himself alone. Around him dropping shots from the enemy's musketry were falling ; but his companions were gone, and the rebels were firing on an empty post. He pulled himself together, and, following in the direction he thought he should take, succeeded in overtaking his comrades, and effecting his escape.

Chapter XVI.

THE INDIAN MUTINY—WINDHAM'S PERIL AT CAWNPORE— SIR COLIN'S DASH TO HIS RELIEF.

LEAVING Lucknow at eleven o'clock on the forenoon of the 27th November 1857, Sir Colin Campbell, carrying with him, besides the scattered remnants of the 32nd Regiment, the whole of the women, children, and treasure rescued from the Residency, set his face for Cawnpore. In all, two thousand helpless human beings—many of them sick, wounded, and dying—had to be borne along by his slender force of three thousand men. The state of Cawnpore caused the Chief much anxiety. General Windham, of good Crimean reputation, had been left behind to guard that "city of horrors," while Sir Colin had gone forward to the relief of Lucknow. Not on one, but on two sides Cawnpore had been beset with danger. Nana Sahib, with many thousands of Sepoys, had been menacing it from Bithoor ever since the time, two months before, that Havelock had found it necessary to fight the rebels there to relieve the pressure on Neill. The Gwalior Contingent, 12,000 to 14,000 strong, had also been hovering about in an aimless sort of way, but with Cawnpore as their probable objective as soon as they got a favourable opportunity of attack. Knowing the hazardous nature of Sir Colin Campbell's task at Lucknow, Windham had been loyally sending forward reinforcements to his chief as promptly as they arrived, and retaining nothing but his own handful of men—thirteen hundred, all told—to meet the closing dangers by which he was himself beset.

During the later days of the struggle at Lucknow, "a thick veil" had suddenly interposed between Sir Colin and Windham. Cawnpore was little more than thirty miles distant, but for days not a whisper had been heard from its garrison. The silence suggested the worst fears, and Sir Colin felt that it behoved him, now that the work at Lucknow was accomplished, to speed quickly back to Cawnpore. Yet, with his huge, lumbering, unwieldy, but precious convoy, progress at the best could not be rapid. The first night the army and convoy camped at the Bunnee Bridge, and in response to anxious inquiries, the commander of the post said he had heard the sound of heavy firing in the direction of Cawnpore both that day and the day before. This confirmed the fears of the Commander-in-Chief, and he felt himself in presence of a momentous crisis. The rebel host had doubtless fallen upon Windham. If that officer was defeated, the bridge of boats over the Ganges would be destroyed, and Campbell, his force, and convoy would be completely isolated in the country of an enemy numbering untold thousands. An overwhelming disaster stared the old General in the face, and he knew that—if not already too late—prompt action alone could save all. " Not a moment was to be lost; the danger was instant." Early astir next morning was Sir Colin, and the day had not long broken when his column was on the march—" eagerly pressing on," says the writer in *Blackwood*, himself a participator in the event, "towards the scene of danger. At every step the sound of a heavy but distant cannonade became more distinct; but mile after mile was passed over, and no news could be obtained. The anxiety and impatience of all became extreme." Before noon a Native, who had been concealed behind a wayside hedge, ran forward and delivered a message to the advanced guard. He handed a small rolled-up letter, written in the Greek character, to the officer in charge, and it was addressed—"Most urgent. To Sir Colin Campbell, or any officer commanding troops on the Lucknow road." It was dated two days earlier, and stated that the fighting at Cawnpore had been most severe, that the enemy—the Gwalior and Bithoor armies united—was very powerful, and that, unless aid arrived, Windham and his troops would require to retire within the entrenchments. The letter concluded by expressing the hope that Sir Colin would lose no time in coming to the assistance of the beleaguered men.

For very life the column had to press on now. Windham once within the entrenchments the city was in the hands of the enemy, and the bridge would be surely destroyed. Every step was quickened, every nerve was strained to reach the Ganges in time. "Loud and louder grew the roar, faster and faster became the march, long and weary was the way, tired and footsore grew the Infantry, death fell on the exhausted wounded with a terrible rapidity—the travel-worn bearers could hardly stagger along beneath their loads—the sick men groaned and died—but still on, on, on was the cry. Salvoes of artillery were fired by the field battery of the advanced guard, in hopes that its sounds might convey to the beleaguered garrison a promise of coming aid." Sir Colin, chafing under the suspense, and eager to know the worst, had hurried on with his Staff, Cavalry, and Artillery, far ahead of the Infantry column and its convoy. From Mungulwar, with the sound of the fierce warfare beyond breaking more loudly on his ear, he dashed forward, still more impatiently, accompanied by his Staff alone. A few miles of hard riding, and

Cawnpore, the scene of a dreadful conflict, was under his eye. "Then the veil which had so long shrouded Windham was rent asunder, and the disaster stood before him in all its deformity."

Windham beleaguered in the entrenchments, was fighting for dear existence. Flames were rising over the city, mingling their lurid glare with the fading red light of the setting sun. The city and a large part of the cantonments had been taken, and all the tents, the stores, and clothing intended for the wounded and the women and children removed from Lucknow were in the enemy's hands. But the Bridge was still intact, and with renewed eagerness Sir Colin dashed forward. It is no part of our plan to detail the circumstances under which Windham had been driven into his present position. Suffice it to say he had gone out to measure his strength against numbers that were overwhelming, and had lost in the battle. 20,000 men, with 40 guns, had driven him back, first from one position, then from another, till, on this November evening, the crowning incident of a great disaster seemed about to befall him and his men. The band of British heroes had almost given up hope of succour, for message after message—intercepted by the enemy—had been sent to Lucknow, and had elicited no response.

But in a moment all was changed. Suddenly the clattering of horses' hoofs was heard on the bridge of boats crossing the Ganges, and in the dusk a few horsemen were seen riding over it at a furious gallop. They spurred quickly up the road leading to the fort, and as they came close under the ramparts it was seen that an old man with grey hair was riding at their head.

"It is Sir Colin!" exultingly cried one of the men, recognising the Commander-in-Chief, and shouting aloud to his companions within.

"The news," says the writer in *Blackwood*, "spread like
wildfire; the men, crowding upon the parapet, sent forth
cheer after cheer. The enemy, surprised at the commotion,
for a few minutes ceased their fire. The old man rode in
through the gate. All felt that the crisis was over—that the
Residency saved would not now be balanced by Cawnpore
lost. When the morning broke, the plain (over the river)
towards Lucknow was white with the tents of the returning
army." A few hours later and there hung over the stream
the smoke of a fierce battle; for the rebels, realising the
importance of the bridge to the returning troops, had directed
their guns towards its destruction. But Sir Colin, who had
already re-crossed to his camp, anticipating the move, had
directed Peel and his Artillery to be ready to meet the
attack. For a time the battle raged undecided; but at
length British pluck and endurance prevailed, and Sir Colin's
troops, under a storm of fire, commenced to cross the bridge.
The 93rd, passing through a whirlwind of shot, shell, and
bullets, moved up towards the position assigned them, not
far from the memorable spot where Wheeler so long withstood the assaults of Nana Sahib's myrmidons. In the
evening the great convoy which Sir Colin had brought with
him was safely carried over the river, and placed in security
in the camp.

And now the British army remained on the defensive for
a few days till Sir Colin felt ready to strike. On the 3rd of
December he sent off his convoy—which had cost him so
much anxiety—strongly escorted, to Allahabad, and then he
set about making his dispositions to attack the rebels, who
were harassing him day and night.

Meanwhile a new force had been added to Adrian Hope's
brigade. The famous Black Watch had come up in hot
haste from Cheemee, where they had been encamped. Not-

withstanding the presence of cholera in their ranks, they had performed a toilsome and trying march with all their old dash and endurance. They had covered seventy-eight miles in three days, and, although tired and exhausted when they reached Cawnpore, were eager to engage the enemy. It greatly delighted the 42nd to meet their old comrades of the 93rd ; and it added to their pleasure when they found themselves again brigaded with the same heroes who had ascended the heights of Alma by their side. In their turn the 93rd gave their compatriots a hearty welcome, for they, too, remembered with pleasure and pride the glorious deeds of the past. The 42nd were a very acceptable accession of strength to Sir Colin. He knew them of old ; knew how brave and true they were in time of trial. One of them who took part in the perils of the campaign writes—" He shook hands with our officers and welcomed us, and said he had some hard work for us to perform."

It was on the 6th of December that the day of "hard work" arrived. By that time Sir Colin had matured his plans, collected his forces, and was prepared to throw down the gage of battle to the thousands of the enemy who had been incessantly harassing his troops. The battle commenced with the roar of Windham's guns. Hope and his Brigade, concealed far away on the left of the British position, were for some time inactive ; but at length they were in motion. Conjointly with Inglis's Brigade, that of Hope marched forward into the plain against the enemy's position, and it was seen that there the full fury of the battle was about to burst. They advanced in columns of companies, and as wave after wave of the dark-plumed Highlanders, their bayonets glittering in the sunlight, moved forward, the sight gave inspiration and hope to all who had time to look. In front, skirmishing in fine order, were the Sikhs and gallant 53rd.

N

The enemy, posted along the line of the canal, and protected by high brick mounds which covered the bridge over the canal, kept up a sweeping fire; but on went the bare-kneed warriors, closing upon them at every step. Riding in front of the 42nd, Lieut.-Colonel Thorold, by whom they were commanded, had his horse shot under him by a round shot, which, sweeping through the line, obtained a human victim in Mark Grant, a private, who was eagerly following his leader. In a moment the brave old Colonel had disentangled himself from his fallen steed, and, springing to his feet, with drawn sword marched in front of his regiment into the thickest of the action. Now grappling with the rebels on the bridge and driving them back, the Highlanders quickly made way for the omnipresent Peel, who with his sailors in a few minutes had a 24-pounder planted on the bridge, and was pouring rounds of grape shot into the ranks of the fleeing foe. Cheering as they went, and with their bagpipes screaming their wild notes of victory, the Highlanders went on with an enthusiastic rush, right into the heart of the enemy's camp. Here they found how little defeat had been expected. Says a writer who was present—"So complete was the surprise, so unexpected was the onslaught, that the chuppaties were found heating upon the fires, the bullocks stood tied behind the hackeries, the sick and wounded were lying in the hospitals, the smith left his forge and the surgeon his ward to fly from the avenging bayonets. Every tent was found exactly as its late owner had sprung from it. Many arose too late, for the conquerors spared none that day —neither the sick man in his weakness, nor the strong man in his strength."

On other parts of the field the victory had been equally complete, and a few hours after the battle opened Cawnpore was relieved, and the highly-trained rebels from Gwalior and

Bithoor were everywhere in flight. "The rebel right, struck by an iron hand, had been shattered into a thousand pieces; their centre, shut up within the town, had been impotent to avert disaster; the camp of the Gwalior contingent, with all the field stores, magazines, and material, was in our power, and the Calpee road, covered with their flying ranks, in possession of our victorious soldiery."

The 93rd were later in the day engaged under General Mansfield in an attack upon the Subadar's tank, where a number of rebels had rallied. This post was on the other side of the city from where they had gained their honours in the morning. It was a stiff one to take, and the Highlanders made a gallant but futile attempt to do so. The enemy, however, evacuated their position in the evening, and drew off towards Bithoor.

The 42nd joined in the pursuit of the rebels, who fled along the Calpee road. For fourteen miles they hotly followed them up. The slaughter was immense, and the strength and courage displayed by the 42nd—many of them young lads, who had joined to fill the gaps made by the Crimean campaign—were commented on by their companions in arms. Many of them had never fired a shot before that morning; yet they engaged in the whole of the arduous fatigues of the day; fought with extraordinary energy; ran after the panic-stricken rebels with the swiftness of deerhounds, then returned in the late evening to camp, without a single man having fallen out from fatigue.

After the Nana's treasure had been dug up from the well at Bithoor—an operation in which both the 42nd and 93rd assisted—Adrian Hope led the Sikhs, 53rd, 93rd, 42nd, a howitzer, 4 light field battery guns, and some engineers to the suspension bridge by which the road from Cawnpore to Futteghur crosses the Kalee river or Nudee. The object of the movement was to secure and repair the bridge, which the rebel troops of the Nawab of Futteghur were endeavouring to destroy. Here the honours of the day were carried off by the 53rd, who, however, snatched them from the very hands of their Highland comrades. In almost all the operations in which the 93rd had been engaged since their arrival in India, the 53rd had been by their side, and a very friendly rivalry existed between the regiments, in the latter of which was a large proportion of Irishmen. At the battle of the Kalee Nudee, in which the attack was hotly and unexpectedly opened by the Nawab's men, the 53rd were pushed across in front to hold the bridge and relieve the pickets, while the headquarters of the 93rd were held in reserve behind. After the conflict had gone on for some time the 93rd were ordered to relieve the 53rd. " But that gallant corps," says Captain Burgoyne, " could not brook the idea of being relieved in advance, and several of their number having been wounded during the day, they were determined also to have their revenge. On seeing the 93rd coming down towards the bridge to relieve them, they could no longer be restrained, but, with a pealing cheer, rose from their cover and dashed into the village (beyond the bridge), just as the enemy, smarting from an Artillery fire, were beginning to retire in good order.

" On went the 53rd, their bugles sounding the advance, and their officers, carried away with the stream, perfectly unable, even had they been willing, to restrain the advance of their men. The 93rd followed quick and eager behind ; Hope Grant, with the Cavalry, moved away to the left, and came crashing on the enemy's flank. Thus pushed in rear by the 53rd, who followed close up, and cut through by the Cavalry, the rebel army—hitherto retiring in perfect order, covered by their light guns—broke and fled in haste in every direction. Gun after gun, standard after standard, fell into our possession, until coming night put an end to the pursuit.

"That advance of the 53rd was a daring act of disobedience, but had its origin in a gallant spirit; and the 93rd, free from all jealousy, could sympathise with the feelings which prompted it."

But Sir Colin, who was present, having gone down in the morning to see how the repair of the bridge was getting on, was terribly angry. He could not brook such a breach of discipline. His eyes blazing with passion, he rode up to the 53rd for the purpose of venting his displeasure. But the gallant fellows were equal to the occasion. As soon as he opened his mouth—led by an adventurous spirit—they shouted with all the strength of their lungs—"Three cheers for the Commander-in-Chief, boys." Again the irate General commenced his scolding, and again his voice was drowned with—"Three cheers for the Commander-in-Chief, boys." Once more he essayed to rebuke, and "Three cheers for the Commander-in-Chief, boys," was the lustily-shouted answer. At length, finding it impossible to obtain a hearing, the stern countenance which he had assumed for the occasion relaxed, and the veteran Chief turned away with a laugh. Sir Colin, it may be added, had just a few minutes before been struck in the stomach by a spent ball, which had strength enough to cut his breath for the moment, but did no further injury.

In this engagement the rebels suffered heavy loss—the Cavalry under Hope Grant inflicting terrible slaughter. As the triumphant heroes were returning to camp, Sir Colin took off his hat to the Lancers and Sikhs, who were proudly waving their captured standards, their lances, and sabres. Thrilling cheers rose from the ranks. The Highlanders, encamped near by, rushed down, "and cheered both the victorious Cavalry and the veteran Chief, waving their bonnets in the air. It was a fair sight," adds Alison, "and reminded one of the old days of chivalry. When Sir Colin rode back through the camp of the Highlanders the enthusiasm of the men exceeded description."

Chapter XVII.

THE INDIAN MUTINY—THE THIRD ATTACK ON LUCKNOW.

THE next event of importance which we are called on to chronicle is the third and final attack on Lucknow. This opened in the early days of March 1858. By this time Sir Colin Campbell had been able to lay his iron hand heavily on the rebel Forces in other parts of the country. He had broken up their organisation, cut off their supplies, captured their guns, and reduced their strongholds In some regions the rebellion was, thanks to the energy and courage of the Commander-in Chief and of the troops under him, almost stamped out, and British authority was once more beginning to assert its power and supremacy.

But the princely capital of Oude still held out. From many quarters the rebels had gathered there; they had thrown up redoubtable defences, and more than a hundred thousand men, nearly three-fourths of whom were trained troops, had assembled to dispute the authority of Britain. For months Outram's posts on the Cawnpore side of the city had been subjected to persistent and determined attacks, but that General had always been able to give more than he got, and so kept intact the positions which Sir Colin had considered it desirable to hold against his return.

As we have said, it was in the early days of March that Sir Colin, having assembled his troops for a supreme effort, appeared once more in front of Lucknow, this time determined to drive the rebels root and branch from the protection of its friendly walls. Inclusive of the troops under Outram, he brought to the attack 20,000 men and 180 guns. With this large force, although numbering only one to five of the enemy, he was able to act on a carefully pre-arranged plan, and leave nothing to chance. The plan was drawn up by Sir Colin and one who has since proved himself an able and distinguished General—Lord Napier of Magdala. It involved a co-operative action on the part of Sir James Outram from the north side of the Goomtee, opposite the city, and also included certain movements on the part of a contingent of troops brought by the Maharajah Jung Bahadoor, a friendly Nepaul Prince, by whom, for political rather than for military reasons, Sir Colin permitted himself to be assisted. Sir Colin did not much care for any help this potentate could afford. He would rather have had an extra hundred Highlanders than an extra thousand of Nepaulese troops. But Lord Canning, the Governor-General, felt it desirable to humour the Maharajah, and Sir Colin had no alternative. His long delay in making the attack on Lucknow was indeed due to the tardy movements of this personage and his contingent, who did not arrive on the scene of action till some days after operations had commenced.

Sir Colin had in the meantime formed a closer relationship with the gallant 93rd. The Queen, on January 19th, had written with her own hand a long letter to the old veteran who was doing such signal service to her cause in India. In the letter Her Majesty said—

" The Queen must give utterance herself to the feelings of pride and satisfaction with which she has learned of the glorious victories which Sir Colin Campbell and the gallant and heroic troops which he has under his command have obtained over the mutineers.

" The manner in which Sir Colin has conducted all these

operations, and his rescue of that devoted band of heroes and heroines at Lucknow, which brought comfort and relief to so many, many anxious hearts, is beyond all praise.

" . . . But Sir Colin must bear one reproof from his Queen, and that is, that he exposes himself too much; his life is most precious, and she entreats that he will neither put himself where his noble spirit would urge him to be, foremost in danger, nor fatigue himself so as to injure his health. . . .

"To all European, as well as Native troops, who have fought so nobly and so gallantly, and amongst whom the Queen is rejoiced to see the 93rd, the Queen wishes Sir Colin to convey the expressions of her great admiration and gratitude."

In forwarding the letter from which the above extracts are taken, the Duke of Cambridge added a short but important note, in which he said—"In consequence of the Colonelcy of the 93rd Highlanders having become vacant by the death of General Parkinson, I have recommended the Queen to remove you to the command of that distinguished and gallant corps, with which you have been so much associated, not only at the present moment in India, but also during the whole of the campaign in the Crimea. I thought such an arrangement would be agreeable to yourself, and I know that it is the highest compliment that Her Majesty could pay to the 93rd Highlanders to see their dear old chief at their head."

The veteran Commander-in-Chief had accepted the honour bestowed upon him, and written to the Queen characterising Her Majesty's letter as the greatest mark of honour that it was in the power of the Royal Lady to bestow. He added that he would not fail to execute the most gracious commands of Her Majesty, and would " convey to the army, and more

particularly the 93rd Regiment, the rem >mbrance of the
Queen."

In the army now drawn up before Lucknow were the
42nd, the 93rd, and the 79th, the old Brigade with
which such wonders were achieved at the Alma. The 79th
had received a hearty greeting from their brethren of the
kilt, and looked fit and ready for the work before them.
They were attached to the force under Outram, and in the
opening operations of the siege acted from the opposite side
of the Goomtee, under Brigadier-General Douglas.

The task before Sir Colin was one of no ordinary importance. If he could crush the rebels here the final suppression
of the mutiny would become a mere matter of time and
detail; but so long as this stronghold was in their possession
so long was Oude unconquered, with a powerful army ready
to commit any act of mutinous aggression. But if the task
was important, the General's plans were equally careful and
elaborate. For months his troops had been moving forward.
Dense battalions and glittering squadrons, long lines of guns
and carriages, the great siege train, and legions of camp
followers had crossed the Ganges, and traversed the plain
leading to the princely city. Such a demonstration India had
never before seen. Gathered between the Alumbagh and the
Dilkoosha was now an army which showed Great Britain
in her pride and her strength. The rebels had now no weary
way-worn handful of men to fight and repel; but a field
force complete in every arm, and led by men of the highest
military skill and most dauntless personal courage.

The mutineers were well posted. As they had increased
in numbers since Sir Colin's last visit, so had they
strengthened and extended their defences, until the great
city was one huge fortification, presenting manifold difficulties to the attacking force. Besides a great outer line of

works reaching from the Goomtee to the Charbagh Bridge, there were three inner lines of the most formidable character, with many cannon in position, and behind the 110,000 or 120,000 men who guarded the works was a hostile population numbering nearly a million souls. It was a great and fair city that was about once more to become the theatre of a terrible conflict—this time the most severe and bloody it had seen. Dr Russell, of the *Times*, who surveyed it some evenings before Sir Colin opened his attack, thus wrote in his *Diary in India:*—"A vision of palaces, minarets, domes azure and golden, cupolas, colonnades, long façades of fair perspective in pillars and columns, terraced roofs—all rising up amid a calm, still ocean of the brightest verdure. Look for miles and miles away, and still the ocean spreads, and the towers of the fairy city gleam in its midst. Spires of gold glitter in the sun, turrets and gilded spheres shine like constellations. There is nothing mean or squalid to be seen. There is a city more vast than Paris, as it seems, and more brilliant, lying before us."

It was over this magnificent city—which had already within the previous few months twice endured the brunt of battle—that in fiercest fury the terrible storm of war was now about to burst. For some days desultory firing had taken place, a man had been picked off here and there by the sharpshooters on both sides; and well-directed cannon shots had more than once caused some little excitement and damage. But it was not till the 9th of March that Sir Colin was fairly ready to open the contest.

Even on that morning, none, save the members of his own staff, knew that the important day had at length arrived. Dr Russell relates that in going through the Highlanders' camp he was asked by the officers "what was up," as they had been ordered to give their men dinner at twelve o'clock.

Shortly after the order was issued. The Martiniere was
to be attacked at two o'clock! And the 42nd, accompanied
by a Punjaub regiment, was selected to lead. This was the
place of honour, and the enthusiasm of the men rose
high. Eagerly the Scots prepared for action of the sternest
kind—glad to relieve the monotony of the dull camp life
which for weeks they had been enduring. The two leading
regiments were to be supported by the 38th, the 53rd, the
gallant Perthshire 90th, and the veteran 93rd.

Just before the action commenced an incident occurred
which is worth remembering, chiefly because of the persons
concerned. Serving, the one in the 42nd and the other in the
93rd, were two brothers, both young men, and the sons of
Cluny Macpherson, a Highland chieftain, than whom no man
in Scotland bore a more justly honoured name. These were
Captain Duncan Macpherson of the Black Watch and
Lieutenant Ewen Macpherson of the 93rd. Duncan was
going first into action with the 42nd, and turning to his
brother he took off his finger-rings, watch, and other trinkets
he wore, and handing them over, said—"Here, Ewen, take
these. If I come out of this all right I'll get them from you;
if I don't they are yours." Both came safely out of the
engagement and out of the war; and, zealously pursuing their
military career, singularly enough each lived to command
the regiment in which he then served—the former being
the distinguished officer who led the Black Watch over
the trenches at Tel-el-Kebir, while at the same date the
latter was Lieutenant-Colonel of the 93rd.

At length the time for action arrived. Just at two P.M.,
sounding above the roar of the cannonade which had been
proceeding for some time, rose the sharp, ringing words
of command, "Forward!" "forward!" "forward!" running
along from right to left; and company by company, the 42nd

moved forward to attack the Martiniere. First went four companies under Major Priestly, then followed the remainder of the regiment (with the exception of No. 6 Company, which was left to guard the camp), led by Colonel Duncan Cameron. The regiment went on in beautiful style until they had got to within two hundred yards of the works, when the pipers striking up " The Campbells are comin'," they broke into the double, and rushed on with bayonets fixed. For some time the rebels in the advance works had kept up a blazing fire ; but as the Highlanders drew near terror seemed to take possession of them, they deserted their rifle pits, crowded into the main passages, and then ran in long white streams towards the Martiniere itself. At increased speed on went the Highlanders, eager to be among the fleeing enemy. For a time the latter summoned fresh courage as, joined by some reinforcements from the second defences, they opened fire on the advancing companies ; but in a few minutes that steadily-approaching, unswerving line proved too strong for their stomachs, and they bolted again. Quickly the Highlanders were in the trenches, leaping into the rifle pits, and in a few minutes more Sir Colin was gratified with the intelligence that they were within the walls of the Martiniere.

The first point in the programme had been carried with a loss that was infinitesimal, and as the 42nd and the Sikhs, still advancing, skirmished away through the suburbs in the direction of the city, firing at and driving back everything they came across, the roar of Peel and Outram's guns, and the bursting and crashing of their heavy missiles in palace and trench, caused the cravens within the city to quake with fear. But the leaders did not yet despair. The Kaiser-Bagh, covered by the Begum Kothie (Begum's Palace), was the rock of their defence, and they still held it with firm grasp.

A former private of the 42nd, who was present with his regiment in the campaign, in some notes of his recollections which he has furnished to us, says—" No. 6 Company, in which I was, had been on outlying picket duty the night before, and took no part in the action of the 9th. We were left to guard the camp, much to the displeasure of our captain and all the rest 'of us. Captain Baird was near by, looking on as the rest of the regiment formed up for action. Captain Macpherson called out to him, 'Come awa', Davie, man.' These two officers always appeared to be great friends. As the regiment advanced, Captain Baird watched its progress with a field glass. Then the fire commenced on both sides, and I heard Baird exclaim— 'Well done, Cluny !' He told us that Nos. 5 and 7 Companies were in the Martiniere, and that the whole regiment was still advancing, and a heavy firing going on. By-and-by the chaplain, the Rev. W. Ross, came up to where one or two of us were standing, and told us there had been good work done. All that night we lay ready to turn out at a moment's notice. Next day a company came up and relieved us, and we marched to where the main body of the regiment was lying. We passed the Martiniere, and I could see plenty of killed rebels lying here and there. But as we approached a big bungalow our hearts were cheered by the sound of the bagpipes playing a foursome reel. When we were halted and dismissed I went into the building, and there were four or five sets up dancing with all their might, Captain Macpherson and Sir David Baird footing it among the rest."

On the 9th Sir Colin had been successful to an extent he had hardly hoped for ; on the 10th he was to measure his strength against the Begum's Palace, and for that duty the 93rd was called to the front.

Chapter XVIII.

THE party selected to storm the Begum Kothie consisted, says Malleson, of those " Companions in glory, the 93rd Highlanders and the 4th Punjaub Rifles. It was indeed fit that to the men who, in the previous November, had stormed the Scundra-Bagh and carried the Shah Nujjif should be entrusted the final difficult enterprise of his (Sir Colin Campbell's) second movement on Lucknow. Fortunate in their splendid discipline, in their tried comradeship, in their confidence in each other, the 4th Punjaub Rifles and the 93rd Highlanders enjoyed the additional privilege of having as their leader one of the noblest men who ever wore the British uniform—the bravest of soldiers, and the most gallant of gentlemen. Those who had the privilege of intimate acquaintance with Adrian Hope will recognise the fitness of the description."

During the night of the 9th of March the 93rd Highlanders bivouacked in a garden surrounded by mud walls, which protected the men from the dropping fire of the enemy. Straight in front of them was the Begum Kothie, a place of enormous strength, and with defences of such a nature that had they been properly known beforehand the General would have hesitated before ordering the advance to attack. Along the front of the series of buildings and courtyards of which it was composed was a mud wall, facing the British positions, and loopholed in many places; the gateways were concealed and protected by strong earthworks, while loopholed parapets surrounded the building. Added to these

obstacles was the most formidable of all—a broad, deep
ditch which ran along the whole front of the position, and
the existence of which was not even guessed at before the
storming party found themselves at its edge. The Palace
was, as a place of such importance demanded, very strongly
garrisoned by Sepoys.

Early on the morning of the 10th big guns were dragged
into the garden, and two batteries of Artillery were shortly
pounding away at the almost invulnerable walls—Outram,
meanwhile, from over the river, blazing and knocking at the
Kaiser-Bagh. The defenders of the building, from behind
their shelters, kept up a rattling musketry fire, which
indicated their determination to offer a spirited resistance.
For the whole day the duel went on, but when night fell
there was no breach visible, and the stormers were not
called into action.[*]

Next day the contest was renewed with better success.
By three o'clock in the afternoon it was evident that the
moment of action was at hand. Breaches had been made in
the breastwork and in the wall of the outer courtyard.
Adrian Hope quickly detailed his men for duty. He told off
the 93rd into two divisions. The right wing, under Colonel
Leith Hay, consisting of the Grenadier Company and Nos. 1,
2, 3, and 4, was to assault and enter by the first breach,
while the remainder of the regiment, under the command
of Lieutenant-Colonel Gordon, would attack the breach on
the flank of the position made by the battery firing from
Banks's Bungalow. At four o'clock the order was passed, the
big guns became silent, and the 93rd at once, in all their
picturesque beauty of movement, emerged from the garden,
moved on towards the position, and without firing a shot, got

---

[*] Colonel Malleson is in error in giving the date of the storming of the Begum
Kothie as the 10th March. There is no reason to doubt it was on the 11th that
the assault was made.

under cover of some ruined buildings. For a few minutes they gathered themselves together for the supreme effort, then the gallant Brigadier gave the signal, and with a cheer the men left the cover, each party dashing at the breach before it. Feeling now that the real time of trial had come, the enemy renewed with redoubled fury their heavy fire, pouring, says Captain Burgoyne, who was present, "a perfect storm of musketry at the advancing columns." They fired too high, and not a man fell; they blazed again and again,

Colonel LEITH HAY, C.B., of Leith Hall, Aberdeenshire, late Colonel, Commanding the 93rd Highlanders.

but although the storm of bullets hissed and whistled over and around the advancing Highlanders, not a man wavered. Still on they went till the ditch was reached, when, surprised by the unexpected obstacle, for a moment they paused. Then Captain Middleton, of Leith Hay's division,

o

leapt into the ditch, and, like a gallant soldier, showed the
way.   He was followed by a few of the Grenadiers, and
immediately the whole of the men were after them.   Leith
Hay himself was among the first across, and having gained,
along with Middleton and a few men, a footing on the other
side, he assisted in helping the rest out of the ditch.   The
left wing, " with equal rapidity and daring," had at the same
time succeeded in getting over this unexpected obstacle, and
now the whole Force made straight for the breaches.

Here the struggle commenced in terrible earnest.   " Every
obstacle," says Captain Burgoyne, "that could be opposed to
the stormers had been prepared by the enemy—every room
door, gallery, or gateway was so obstructed and barricaded
that only a single man could pass at a time.   Almost every
window or opening that could afford the slightest shelter was
occupied by an enemy, and in threading their way through
the narrow passages and doorways, our men were exposed to
unseen foes."   But nothing would keep back those resolute
men.   Sir Colin had made no mistake in delegating to them
this work of superlative importance.   They were proud of
the distinction, and were determined to emerge from the
Begum Kothie only as conquerors covered with the glory of a
great victory.   "Individual valour," says Malleson, "inspired
by a determination to conquer, was not to be withstood."
Barrier after barrier was passed, the Highlanders grappling
with their foes, and overturning them at every point, until
at length they emerged into the first square of the building,
and there beheld the mutineers in large numbers ready, and
apparently willing, for battle.

There was no pause, no halting hesitation of a moment.
The men saw their enemy in front, and, obeying the sharp
and ready words of command, dashed forward.   They neither
thought of the enemy's greater numbers nor of their

advantages of position. Instantly rifle and bayonet were at work, and the battle raged hand to hand. This was no conflict of a few minutes. For two whole hours it continued —the Highlanders, courageously supported by the Punjaubees, performing prodigies of valour. Above the roar of the battle was sounding the wild war notes of the bagpipes—sweetest music in a Highland soldier's ear—for John Macleod, the Pipe-Major of the 93rd, remembered well his duty in the turmoil. He had been among the first to force his way through the breach, and no sooner was he within the building than he began to encourage the men by vigorously playing his pipes. The more hot and deadly the battle became the more high-strung became the piper's feelings, and the more loudly did the bagpipes peal and scream—John standing the while in positions perfectly exposed to the fire of the enemy, to whom doubtless he appeared as some unearthly visitant.

The battle was a dire repetition of that within the court-yard of the Secundra-Bagh. There was no quarter asked ; there was none given. It was disablement or death to all that could be reached. From room to room, from court to court the rebels were driven. There was no escape from the furious Highlanders—neither in the hazard of combat nor in the eagerly sought chance of concealment. Shot and bayonet did their work surely and remorselessly, and when at length, taking refuge in flight, the last survivor of the garrison evacuated the place he left dead behind him nearly nine hundred of his comrades—bearing testimony to the terrible struggle which had raged within the palace. Many wounded were also carried off. It was, in the language of the Commander-in-Chief, " the sternest struggle which had occurred during the siege."

Officers and men had fought with equal valour, many of the former engaging in personal combat against desperate

odds. One especially distinguished himself — Adjutant
William M'Bean, who will be remembered as the gallant
officer who, under fire, in the preceding November, hoisted
the flag on the Shah Nujjif as the signal to Outram that the
place was captured. Than M'Bean there was no better
known or better liked man in the 93rd. He had already
been twenty-three years in the regiment. Joining when a
barefooted lad as a private at Inverness, he had worked his
way up to his present position, and, we may add, lived not
only to command the gallant regiment to which he belonged,
but to bear the rank of a Major-General in Her Majesty's
Army. This 11th of March, however, was a red letter day
in his remarkable career. Again and again he was met by
desperate odds, but, fighting like a lion, with his own hand
he killed eleven of the enemy, a feat which secured for him
the proudest of all his honours—the Victoria Cross. Though
come of a fighting race, and brave as a lion, M'Bean was
a simple-minded man. "When," says Surgeon-General
Munro, in his "Reminiscences of Military Service with the
93rd Highlanders," "a regimental parade was held for the
purpose of presenting his well-earned cross, as the General,
Sir R. Garrett, pinned the decoration to his breast, he
addressed him in the following words :—

" 'This cross has been awarded to you for the conspicuous
gallantry you displayed at the assault of the enemy's position
at Lucknow, on which occasion you killed eleven of the
enemy by whom you were surrounded. And a good day's
work it was, sir.'

" 'Tuts,' said my gallant and simple friend, quite forgetting that he was on parade, and perhaps a little piqued at
his performance being spoken of as a day's work, ' Tuts, it
didna tak' me twunty minutes.'

" These are the very words he used, spoken in his own

braid Scotch too. I who tell the story heard them, for I was standing by his side."

In all his campaigning M'Bean never received a scratch.

Major-General WILLIAM M'BEAN, V.C.

A sad death was that of poor Captain Charles William Macdonald, the son of Sir John Macdonald, of the 92nd. He had gone through the work of the Crimea without a scratch ; and the same good fortune had followed him throughout the present campaign. But it altogether deserted him on this fatal day. Before the assault he had been severely wounded by a shell splinter on the sword arm. The brave young officer refused to retire, however, and went on with his men. As he entered the breach at the head of his company he was again wounded, a shot passing through his thigh. He fell disabled, and was lifted for removal to the surgeon. As he was being carried back, a bullet passed through his neck and killed him in the men's hands. All the gallant fellows could do was to return to the front and avenge his loss.

Considering the nature of the fighting and the slaughter of the enemy, the casualties of the 93rd were slight—two officers and thirteen men killed, and two officers and forty-five men wounded. One of those who fell during this assault was a remarkable man, who, although not a member of the 93rd, had shared their dangers, and met his fate. This was Hodson, of Hodson's Horse, one of the most striking individuals engaged in the suppression of the mutiny. It was he who had captured the King of Delhi, and in cold blood shot his sons, thus retaliating in a manner that could not be mistaken for the cruelties that had been perpetrated on his countrymen and women. Hodson was a warrior born. His delight was to be in the roar and tumult of battle. With Burnaby he would have exclaimed—"There is no sport like war, my boys !" He was reckless of his own life ; unscrupulous in the robbing of it from others. " He joyed," says Malleson, " in the life of camps, and revelled in the clang of arms. His music was the call of the trumpet ; the battlefield his ballroom. He would have been at home in the camp of Wallenstein, in the sack of Magdeburg. In him human suffering awoke no feeling, the shedding of blood cost

him no pang, the taking of life brought him no remorse."
Yet he did good work in the Queen's behalf. He has been
blamed for his murder of the sons of the King of Delhi, but
he regarded them as ruffians, whom to spare was treason.
This bloodthirsty giant had forced his way through the
93rd—his nearest companion being Robert Napier (now Lord
Napier of Magdala). The two had got separated in the *mêlée,*
and Hodson went hunting for the enemy wherever he could
be found. At last he came upon a group, and with all his
furious instincts ablaze dashed upon them. But courage alone
is no match for powder and lead; in a moment a volley was
discharged at him, and he fell mortally wounded. The Highlanders saw the incident, and were quick to avenge Hodson's
death. They closed with the foes, and left them not till
every rebel lay dead at their feet. Sir Colin Campbell, in
touching words, communicated the loss to Hodson's widow.

Adrian Hope himself had led one of the storming parties.
He had been pushed in through a window by his men, and
landed headlong among a crowd of Sepoys who filled the
room. They bolted at once on seeing the huge red Celt
gather himself together, and with sword in one hand and
pistol in the other, prepare to give them battle. But Hope
was not content with this mere demonstration of his valour.
With characteristic daring he engaged in the struggle of the
day, and shared its dangers; his conduct, in the words of the
Commander-in-Chief, meriting " special notice."

The gaining of the Begum Kothie was a point of the first
importance. The enemy were not yet subdued, but their
chief stronghold was in our hands, and the rest of the
positions could be taken in detail. The 93rd had fought
with conspicuous bravery. In his general order issued the
same evening, Brigadier Lugard said :—

" The Brigadier-General has shared in many hard fought

actions during his service, but on no occasion has he witnessed a more noble and determined advance than was made by the 93rd this day."

Sir Colin Campbell also in his despatch bore testimony to the brilliant conduct of the Sutherland men. His words were :—

" The manner in which the 93rd Regiment flung itself into the Begum Kothie, followed by the 4th Sikhs, and supported by the 42nd, was magnificent."

This ended the hardest of the work for the Highland regiments engaged in the storming operations. Two companies of the 42nd had a stern engagement in the vicinity of Banks's Bungalow. They fought with all their dash and *sangfroid*, and the enemy could not stand before them. During the succeeding few days the contest raged with never-varying success to the British arms, and finally by the 20th of March the rebels were driven out, and Lucknow was captured.

But even before the victory was completed there had commenced a scene of plunder and pillage which cannot be described. Dr Russell saw the men in the full intoxication of their looting madness, and he pronounces the scene as beyond his power to paint. " It was," he says, " one of the strangest and most distressing sights that could be seen ; but it was also most exciting. Discipline may hold soldiers together till the fight is won ; but it assuredly does not exist for a moment after an assault has been delivered or a storm has taken place. . . . Lying around the orange groves are dead and dying Sepoys, and the white statues are reddened with blood. Leaning against a smiling Venus is a British soldier shot through the neck, gasping, and at every gasp bleeding to death. Here and there officers are running to and fro after their men, persuading or threatening in vain. From the broken portals issue soldiers laden with loot or plunder. Shawls, rich tapestry, gold and silver brocade, caskets of jewels, arms, and splendid dresses. The men are wild with fury and the lust for gold—literally drunk with plunder. Some are busy gouging out the precious stones from the stems of pipes, from saddle cloths, or the hilts of swords, or the butts of pistols and firearms, others swathe their bodies in stuffs crusted with precious metals and gems.

. . One fellow having burst open a leaden-looking lid, which was in reality of solid silver, drew out an armlet of emeralds and diamonds and pearls, so large that I really believed they were not real stones." In the Kaiser-Bagh "the soldiers had broken up several of the store rooms and pitched the contents into the court, which was lumbered with cases, with embroidered cloths, gold and silver brocade, silver vessels, arms, banners, drums, shawls, scarfs, musical instruments, mirrors, pictures, books, accounts, medicine bottles, gorgeous standards, shields, spears, and a heap of things which would make this sheet of paper like a catalogue of a broker's sale. Through these moved the men, wild with excitement, 'drunk with plunder.' I had often heard the phrase, but never saw the thing itself before. They smashed to pieces guns and pistols to get at the gold mountings and the stones set in the stocks. They burned in a fire, which they had made in the centre of the court, brocade and embroidered shawls for the sake of the gold and silver. China, glass, and jade they dashed to pieces in pure wantonness ; pictures they ripped up or tossed on the flames ; furniture shared the same fate."

But we drop the curtain on this distasteful result of a hard won victory. Excuses could be found for the poor fellows who indulged in such mad plunder ; but it may be better to leave the subject undiscussed. The part taken in the siege by the 79th we shall briefly describe in our next chapter.

Chapter XIX.

A S we have said, the operations of the 79th at the Siege of Lucknow were distinct from those engaged in by the two other Highland regiments present. The 79th were included in the Force under General Outram co-operating with the main body from the opposite side of the river Goomtee. The details of these sectional operations—performed under Brigadier Douglas—are very meagre, and history has not done justice to the men who so gallantly carried out their General's intentions. Mr Keltie, in his "History of the Highland Regiments," complains of the want of information regarding the work of the 79th in India, alike in the Regimental Records and elsewhere. Since that work was written, however, the "Life of Outram," by General Goldsmid, has been published, and contains a memorandum from Outram, which gives concise particulars of his operations from the 6th to the 19th of March. By this memorandum, as giving the fullest information available, we shall be guided as t ; the facts which follow. On the 9th of March—the same day as that on which the Martiniere was captured by Sir Colin—Outram had determined to try his strength on the enemy's position. In this operation the 79th were engaged, and in gallant style assisted to carry the Chuker Kothie, or Yellow House, in splendid style. As this was the key of the enemy's first position, the success carried with it important results in favour of the main attack. Next day—that on which Sir Colin's guns were occupied pounding at the Begum Kothie—was chiefly spent by Outram in strengthening his position ; but an exciting event occurred which varied the monotony of the proceedings. In front of a piquet, held by the 79th in the suburbs, the enemy suddenly showed in considerable force, and briskly advanced to the attack. They poured in a heavy fire, but the Highlanders stood their ground, and fought them hard. In the end the result, which at first might well have been doubtful, was a splendid victory for the 79th, which not only repulsed the attack, but inflicted a heavy loss on the too adventurous rebels.

On the day following, the 79th took part in an important movement. They formed part of the first of two divisions with which Sir James projected an attack on the stone bridge across the Goomtee, which the enemy held in strong force. and by which, when beaten in front, they might be able to escape. In their progress a camp of the enemy's Cavalry was surprised and captured. The General, however, on reaching the point of his operations, found the bridge too well guarded by the enemy, and too much under the sweep of their fire, to be able to hold it although he should capture it. He therefore retired, destroying quantities of the enemy's ammunition by the way.

On the 16th of March, acting under instructions from the Commander-in-Chief, Outram crossed the Goomtee and entered the city, having been preceded by the 79th and other regiments under Brigadier Douglas. Some fighting ensued, the 79th being put in possession of the Immambara. Next day a sad accident occurred, from which it is probable, though we cannot be certain, that some of the 79th suffered. As Outram, with a wing of the Highlanders, a company of Middleton's Field Battery, a wing of the 20th, a wing of the

23rd, and some Native troops, was marching to occupy a position, he reached the Juma Musjid, where, in a courtyard, nine cartloads of gunpowder were found. As these impeded the progress of his troops, the General ordered their destruction. This was being done, under the supervision of the Engineers, when, through some accidental cause, the powder ignited, causing the death of two officers and thirty men, and injuring many of the working party.

Two companies of the 79th then advanced to secure the post in question, which they did without loss, the enemy retiring on their appearance, and leaving to be captured an iron and brass gun, an ammunition waggon, and several small guns, all in position, and which might have been used with disastrous effect on the 79th.

Early on the morning of the 19th March, Outram proceeded to attack the Moosa Bagh—the first severe conflict taking place between the enemy posted in Ali Nuki Khan's house, and two companies of the 79th, led by Lieut. Evereth. The Highlanders, however, drove out the enemy in gallant style, and took possession of the house. Later in the action the 79th were sent forward in skirmishing order against the enemy, who showed in great strength on the road. Aided by the guns of Captain Middleton's battery, the skirmishers drove the enemy back, and a flank movement of Outram's Lancers served to complete their utter rout. The Moosa Bagh was then taken.

Such is a mere synopsis of the work of the 79th in the final attack on Lucknow. It gives little idea of the brilliant movements performed by the men of the Cameron Highlanders—their steady discipline, their unshaken courage in face of superior numbers, their disregard of fatigue and hardships. But we can understand that, led by a soldier of Outram's calibre, the utmost possible would be demanded of them. He spared not himself, and those who followed him were quickly made familiar with all the hazards of war. But the brave regiment was equal to all that was required, and won the admiration of its able and chivalrous leader. The loss sustained by the 79th during the whole operations at Lucknow was 7 non-commissioned officers and privates killed ; and 2 officers—Brevet-Major Miller and Ensign Haine — and 21 non-commissioned officers and privates wounded.

Recurring to the 93rd, one incident yet requires to be mentioned. On the 27th of March—the last day of the fighting at Lucknow—No. 6 company, under Captain Burroughs, was on guard at the Burra Durree gateway, when it was reported that some Sepoys held possession of a house near the post, and were firing at all passers by. Captain Burroughs at once started with a party to dislodge them, and having gained the top of the flat-roofed house occupied by the Sepoys, he was making arrangements to dislodge them, when he saw a puff of smoke beneath him. Instantly suspecting an explosion, Burroughs sprang down the stairs, but too late to escape. The staircase was blown from under him, a brick struck his right leg, breaking it. As he fell the leg was broken again, and he was covered by the falling wall of the building. In a sadly bruised and injured condition the unfortunate officer was extricated and removed to the Dilkoosha, where he was put under chloroform, and his twice broken limb set. The explanation of the explosion was that a party of another regiment, bent on the same errand as the party under Burroughs—but each knowing nothing of the other—had got within the building, and resolved to clear out the enemy by bringing it down about their ears. They were entirely successful, and Burroughs in due time perfectly recovered.

The 7th of April saw the three Highland regiments joined to Walpole's Field Force, and brigaded together under Adrian Hope, ready to march into Rohilcund. On the 9th the march commenced, and was continued day by day until the Fort of Rooyah was reached. Here the rebel leader refused to surrender, and an attack was made. The fort was situated in a dense jungle, which almost completely hid it from view. Four companies of the 42nd were sent forward to cover the guns and reconnoitre. This advance proved no child's play. From tree tops, from the top and loopholed walls of the fort, and from many points of vantage, the rebels fired upon the advancing Highlanders. Their fire told well, and many of the Highlanders fell.

At this stage a most lamentable event occurred; one affecting not only the troops engaged, but the whole British army in India. Brigadier Adrian Hope, as was his wont, scorning danger, went forward with the leading detachment of the 42nd, eager to find out the arrangement of the fort, and to discover if there was any way of entering. As he showed himself over a bank he was shot just over the collar-bone in the left side, and died in about ten minutes afterwards. His orderly—young Lieut. Butter of Faskally—was beside him when he fell. Butter saw only some blood on his trousers, and said, "I hope you are not much hurt?" His answer was, "It's all over with me." The Lieutenant then ran and got some water, and would have gone off for a doctor, but Hope called him back. He shook the young man by the hand, and said, "Good-bye, Archie; remember me to all friends." These were his last words; he shut his eyes, and with the sound of battle in his ears, passed from life into death. Lieutenants Bramley and Douglas at the same moment received their death wounds.

The terrible nature of their loss stunned the men of the 42nd, who were without orders to advance or retire, occupying a very exposed position, and were losing men without being able to inflict loss upon the enemy. The order to assault was never given—had Hope lived, Burgoyne has no doubt it would have been—and after remaining exposed to the fire of the enemy for six hours, the 42nd and Punjaubees slowly retired.

This was no brilliant dashing deed of arms, but it was one which severely tried the stamina of the men of the 42nd ; and while all came well out of the ordeal, several conspicuously distinguished themselves. John Simpson, the Quartermaster-Sergeant of the regiment, when the troops had fallen back, heard that one of his officers was left behind in the ditch. The brave fellow at once ran forward alone in the face of a withering fire, and carried Lieutenant Douglas back to the lines. Nor did this feat end his heroism. He returned to the ditch a second time and brought away a private soldier, who like his officer had been so severely wounded as to be unable to follow his retiring comrades. Simpson died at St Martins, near Perth, of paralysis, only so recently as 11th October 1884, at the age of fifty-nine. While in India he was one of the hardened veterans of the regiment who had served in the Crimean campaign of 1854-55. He had been present at the battles of the Alma and Balaclava, took part in the expedition to Kertch and Yenikale, suffered the hardships of the siege of Sebastopol, and shared the dangers of the attacks of the 18th of June and the 8th of September. He served with the regiment throughout its share of the work of suppressing the Indian Mutiny, including, as we have seen, the attack on the rebels at Cawnpore, and subsequent actions up to the Rooyah affair, where, by carrying out Douglas and the wounded private

from the very teeth of the enemy, he earned undyir.g
distinction. For this deed of generous heroism he was

MAJOR SIMPSON, V.C.

presented with the Victoria Cross, and personally thanked by
the Commander-in-Chief. He was decorated with the
Crimean, Turkish, and Indian Mutiny medals, and had
clasps for Alma, Balaclava, Sebastopol, and Lucknow, as well
as his Victoria Cross. He obtained a commission as Quartermaster of the 42nd Regiment at India on 7th October 1859.
On the formation of the Brigade Depôts he was appointed
Quartermaster of the 55th Brigade Depôt at Fort-George on
20th July 1873. He was transferred to the 58th Brigade
Depôt at Stirling in November 1874 ; and on 1st April
1879 to the Perth Militia. He was respected and admired
by all who knew him, as one whose services constituted
a noble page in the history of the distinguished regiment.
He retired with the rank of Major in 1883. Major
Simpson was awarded the good service pension of £50 only
on June 1884, but, unfortunately, he enjoyed the emolument
but a few months. The Major was subsequently gazetted
Quartermaster of the 2nd Perth Highland Volunteers, which
corps, with the 3rd Battalion, buried him with military
honours.

Private James Davis, of the 42nd (a Perth man), distinguished himself in the same way as Simpson, by carrying
in the body of Lieutenant Bramley from under the very
walls of the fort. "Right worthily," writes General Shadwell, "did the grant of the Victoria Cross commemorate
these noble deeds." Yet another gallant veteran soldier of
the 42nd performed a valorous deed on this occasion. The
body of Lieutenant Willoughby lay on the top of the glacis,
and an officer of the Punjaub Rifles went to carry it in.
Lance-Corporal Alexander Thompson, already a medalled and
tried soldier, volunteered to accompany him, and along with
Private Edward Spence, a brave lad who fell during the
attempt, dashed forward in the face of the enemy's incessant

P

fire. Notwithstanding the desperate nature of the attempt, it
was quite successful. For this deed of heroism Thompson
was also awarded the Victoria Cross, and lived for many
years to bear it with honour. He was discharged full
sergeant in 1863, and died in Perth on March 29th, 1880, in
his 57th year.

Sergeant ALEXANDER THOMPSON, V.C.

(Received the Victoria Cross for conspicuous valour at Rooyah.)

At nightfall the enemy evacuated the fort, apparently
without having sustained any serious loss. Thus the object
of the engagement was accomplished; but the result could
not lighten the gloom which the death of Hope had thrown
upon all. He was at the time of his death second Lieutenant-
Colonel of the 93rd; was a soldier of great promise, and
beloved as a brother by all the officers and men. When

Sir Colin Campbell heard of the sad event, he was deeply moved. In his despatch he wrote—" The death of this most distinguished and gallant officer causes the deepest grief to the Commander-in-Chief. Still young in years, he had risen to high command, and by his undaunted courage, combined as it was with extreme kindness and charm of manner, he had secured the confidence of his Brigade in no ordinary degree." Thus was closed the career of one who had already seen much service, performed many brave deeds, and who had before him a great and promising future. His death caused a vast amount of discontent among the officers of the Highland regiments. They were furious with Walpole, who had, they alleged, bungled the assault, and sacrificed life for no purpose. The men, too, were wild, and when Hope was buried their officers were afraid of "mutiny, or worse." They felt angry and ashamed at having to go back from a foe they had only menaced, and against whom no opportunity had been given of measuring their strength.

The loss of the 42nd, who had borne the brunt of the enemy's fire, was severe. Two officers, one sergeant, and six privates were killed, and three sergeants and thirty-four privates wounded, some of them very severely. The 93rd had only five men wounded. Hope's death permitted the advancement of Colonel Leith-Hay to the rank of Brigadier, and Captain Middleton for a time assumed the command of the 93rd.

Nothing of importance occurred to the little Force until the 27th of April, when, after it had been considerably reinforced by troops of all arms, Sir Colin Campbell himself took command, and directed his advance towards Bareilly, where the enemy still held out in great strength. The march was long and very trying. The weather was now getting extremely hot, and disease was prevalent in the British ranks.

Cholera, dysentery, fever, and sunstroke were doing deadly work, and many a poor fellow succumbed to these who had braved every form of the enemy's resistance. On the first day of the long arduous march Dr Russell, of the *Times*, stood by Sir Colin as his "pet Highlanders marched past." As the old man looked admiringly on their sprightly yet solid ranks, he remarked—"The difficulty with these troops, Dr Russell, is to keep them back; that's the danger with them. They will get too far forward." "The Highlanders," adds Russell in his "Diary," "are very proud of Sir Colin, and he is proud of them."

On the march the men were depressed by the intelligence of the death of General Penn, and really saddened by the announcement that Sir William Peel had died at Cawnpore. Every soldier loved brave Captain Peel, of the *Shannon*, who with his band of sailors had done such signal service throughout the campaign. He had received his knighthood only a few days before his death.

The morning of the 5th of May saw the British Force in front of Bareilly. Shortly after six o'clock the advance was commenced, and not long after the enemy's outposts were found. The strength of the British column was 7637 men. The 78th, which had accompanied Walpole's column from Lucknow, was the foremost of the Infantry, and was followed by the heavy batteries, flanked right and left by the 42nd and 93rd. In the next line was the 79th. The battle was opened by a fierce cannonade, which drove the enemy back from the bridge crossing the Nullah in front. They immediately occupied secure positions among the trees and ruined houses in the cantonments. It was slow work driving them out—the British fighting under the rays of a sun which almost scorched them.

About eleven o'clock an unexpected phase of the battle

developed itself. This was nothing less than an attempt on
the part of the enemy to turn the British flank. It was
made by a strong party of Ghazees, and was afterwards
described by Sir Colin as " the most determined effort he had
seen made during this war."

Chapter XX.

IN the operations which followed the opening cannonade at Bareilly, the 4th Punjaub Rifles—who had so distinguished themselves alongside of the 93rd during the preceding months—had been pushed forward to occupy the old Cavalry lines. They had just taken possession of the lines, and were still in broken order, when, availing themselves of the opportune moment, the band of Ghazees—Mussulman fanatics—rushed forward. "Brandishing their swords, with heads low, and uttering the wild cry of their faith, they fell with great impetuosity upon the Sikhs, and drove them back upon the 42nd Highlanders!" Sir Colin, with ready eye, had seen the movement, been quick to detect its object, and had sent forward the 42nd to check the fanatical rush. In a moment friend and foe were rushing pell-mell upon the Highland line. The Punjaubees were in full flight, and the Ghazees were among them slaughtering right and left with their sharp gleaming tulwars. For a moment the Highlanders hesitated. They were afraid to fire lest they should shoot down the struggling Punjaubees. Then as by magic their decision was taken, for behind them the voice of their old chief was heard shouting—

"Fire away, men! shoot them down, every man-jack of them."

The presence of Sir Colin had the desired effect. In an instant their rifles were levelled, and they opened fire upon the maddened enemy. Too late to repel the charge, however; but not too late to meet it with that resolute strength which the British soldier ever shows when the conflict becomes close and deadly.

"Be steady, men, and trust to the bayonet!" Sir Colin had again shouted, and the men handled their steel with deadly effect. In the first rush some of the enemy had worked round to the rear of the Highlanders' line, and fallen upon them from behind—an advantage well calculated to effect disastrously the men attacked. But the 42nd proved themselves to be made of the sternest, most genuine stuff, and never for a moment wavered. The rear rank faced about and tackled the foe. In the struggle heroic deeds were done. Hand to hand the conflict raged, and was terrible while it lasted—the clashing and tumult of the battle being rendered more hideous by the wild cries of the fierce Ghazees. In a few minutes all was over—every "man-jack" of the fanatics had paid the penalty of his temerity. In rear of the line during the battle four of them seized Colonel Alexander Cameron, and attempted to drag him off his horse and despatch him as he fell. Hard as he struggled, Cameron, whose revolver was in his holster, and whose sword had dropped from its scabbard, would inevitably have been killed, had not Colour-Sergeant Gardiner rushed to his assistance. In a moment Gardiner had his bayonet through a couple of them, and the rest were finally despatched, Colonel Cameron escaping from the *mêlée* with nothing more severe than a cut on the wrist. Gardiner received, as he well deserved, the Victoria Cross for his heroic deed. General Walpole was in extreme danger for a time, and would have lost his life but for the promptitude with which the men of the 42nd handled their bayonets. Sir Colin himself had a narrow escape. As he was riding in rear of the 42nd a quasi-dead Ghazee, with

tulwar in hand, caught the Chief's eye gathering himself
for a spring   "Bayonet that man," cried Sir Colin to a
Highlander, reining up on the instant.   The soldier lunged
his bayonet at the fanatic's body, but the bayonet would not
pierce the thick woollen tunic he wore, and rising to his feet
the Ghazee continued the struggle till a Sikh lopped his head
off with one sweeping blow of his sword.   In this affair only
one Highlander, a private, was killed ; but two officers, one
sergeant, and twelve men were wounded.

An incident occurred about this time which is well worth
relating.   Dr Russell, the famous war correspondent of the
Times, accompanied the column.   He had been some days
before kicked by his horse, and the results of the accident
and terrible heat were such that he was put on the sick list,
and carried on a dooly along with the sick and wounded.
During the march Russell had chafed and pined at his misfortune.   "Looking out of my portable bedstead," he says,
"I could see nothing but legs of men, horses, camels, and
elephants moving past in the dust.   The trees were scanty
by the roadside.   There was no friendly shade to afford
the smallest shelter from the blazing sun.   I had all the
sensations of a man who is smothering in a mud bath."
After speaking of the many halts, and the terrible agony
many of the men were suffering—some of them dying in the
terrible heat—he continues—"The crush on the road became
more tremendous.   The guns were beginning to move, every
moment a rude shock was given to the dooly, so I told
the bearers to lift me and carry me to a small tope in the
field on my left, which seemed to be a quarter of a mile
away, and to be certain to give us shade."   But it turned
out that the tope was by no means so umbrageous as the
sufferer thought ; yet he was content to rest while the
bearers sat down to smoke, or perhaps sleep.   Then he goes

on—"Around us just now there was no sign of the British troops in front. They had dipped down into ravines, or were at the other side of the road. Here and there were clouds of dust which marked the course of cavalry. Behind us were the columns of the rearguard and of the baggage. But the camp followers were scattered all over the plains, and the scene looked peaceful as a hop-gathering. There is a sun, indeed, which tells us we are not in Kent. In great pain from angry leech-bites and blisters, I had removed every particle of clothing except my shirt, and lay panting in the dooly. Half-an-hour or so had passed away in a dreamy, pea-soupy kind of existence. I had ceased to wonder why anything was not done. Suddenly there was a little explosion of musketry in our front. I leaned out of my dooly, and saw a long line of Highlanders, who seemed as if they were practising independent file firing in a parade ground, looking in the distance very cool and quiet and firm ; but what they were firing at I in vain endeavoured to ascertain. A few Native troops seemed to be moving about in front of them. As suddenly as it had begun the firing died out once more. . . . A long pause took place. I looked once or twice towards the road to see if there were any symptoms of an advance. Then I sank to sleep. I know not what my dreams were, but well I remember the waking. . . . There was a confused clamour of shrieks and shouting in my ear. My dooly was raised from the ground, and then let fall violently. I heard my bearers shouting ' Sowar ! Sowar !' I saw them flying with terror in their faces. All the camp followers in wild confusion were rushing for the road. It was a veritable stampede of men and animals. Elephants were trumpeting shrilly as they thundered over the fields. Camels slung along at their utmost joggling stride, horse and ponies, women and children, were all pouring in a stream, which converged and tossed in heaps of white as it neared the road—an awful panic. And, heavens above! within a few hundred yards of us, sweeping on like the wind, rushed a great billow of white Sowars (rebel horsemen), their sabres flashing in the sun, the roar of their voices, the thunder of their horses, filling and shaking the air. As they came on, camp followers fell with cleft skulls and bleeding wounds upon the field; the left wing of the wild Cavalry was coming straight for the tope in which we lay. The eye takes in at a glance what the tongue cannot tell or hand write in an hour. Here was, it appeared, an inglorious and miserable death swooping down on us in the heart of that yelling crowd. At that instant my faithful Syce, with drops of sweat rolling down his black face, ran towards me, dragging my unwilling and plunging horse towards the litter, and shouting to me as if in the greatest affliction. I could scarcely move in the dooly. I don't know how I ever managed to do it, but by the help of poor Ramdeen I got into the saddle. It felt like a plate of red-hot iron, all the flesh of the blistered thigh rolled off in a piece on the flap; the leech-bites burst out afresh; the stirrup-irons seemed like blazing coals; death itself could not be more full of pain. I had nothing on but my shirt. Feet and legs naked—head uncovered—with Ramdeen holding on by one stirrup-leather, whilst, with wild cries, he urged on the horse, and struck him over the flanks with a long strip of thorn—I flew across the plain under that awful sun. I was in a ruck of animals soon, and gave up all chance of life as a troop of Sowars dashed in among them. Ramdeen gave a loud cry, with a look of terror over his shoulder, and leaving the stirrup-leather disappeared. I followed the direction of his glance, and saw a black-headed scoundrel ahead of three Sowars, who was coming right at me. I had neither sword nor pistol.

Just at the moment a poor wretch of a camel-driver, leading his beast by the nose-string, rushed right across me, and, seeing the Sowar so close, dashed under his camel's belly. Quick as thought the Sowar reined his horse right round the other side of the camel, and as the man rose I saw the flash of the tulwar falling on his head like a stroke of lightning. It cleft through both his hands, which he had crossed on his head, and, with a feeble gurgle of 'Ram! Ram!' the camel-driver fell close beside me, with his skull split to the nose. I felt my time was come. My naked heels could make no impression on my panting horse. I saw, indeed, a cloud of dust and a body of men advancing from the road, but just at that moment a pain so keen shot through my head that my eyes flashed fire. My senses did not leave me. I knew quite well I was cut down, and put my hand up to my head, but there was no pain; for a moment a pleasant dream of home came across me; I thought I was in the hunting field, that the heart of the pack was all around me, but I could not hold in my horse; my eyes swam, and I remember no more than that I had as it were a delicious plunge into a deep, cool lake, in which I sank deep and deep, till the gurgling waters rushed into my lungs and stifled me."

This is all Dr Russell remembers of his exciting adventure. When he recovered consciousness, after being long insensible, he was in a dooly by the roadside, suffering from violent spasms in the lungs, which caused him to spit up much mucus and blood. He had been caught, just when falling from his horse, by Alexander Robb, a private of the 42nd, who has furnished us with the following narrative of the incident :—

"I was," says Robb, "on the convalescent list, and was not engaged in the battle of Bareilly. I was lying with the rearguard. As the line under the Commander-in-Chief was advancing under the cantonments of Bareilly a heavy firing commenced, and I saw my regiment in line firing and advancing. Suddenly a stampede got up, such as I never saw the like of—camels, elephants, dooly-bearers, camp followers, all mixed up in the general hubbub. At first I could not make out what it meant, but looking to the right I saw some Sowars galloping down on the flank. My attention was quickly attracted to the figure of a man riding towards me on a pony, and with nothing on but his shirt, and bareheaded. This was about twelve o'clock, and the sun was burning hot. He was just in the act of falling off, when I caught him in my arms. I said—'Sir, what regiment do you belong to?' He replied, very faintly—'To no regiment.' Then said I—'What the devil are you doing here?' He told me he was Dr Russell, and his palanquin coming up soon after, I helped him into it. Within it were all his writing materials. I told them to take him to the doctors in the rear, and never heard any more about the matter till I was discharged and settled in Dundee, when I happened to tell the Rev. George Gilfillan of the incident. Mr Gilfillan asked if Dr Russell never wrote or inquired after me. I said I didn't think he ever had the chance, for he was long ill afterwards, and had to advance with the Commander-in-Chief. As for myself, I had never thought more about it. Gilfillan had, I suppose, sometime after seen Dr Russell, for I got a letter from the Doctor, through Mr Gilfillan, asking me a number of questions, which I answered. I afterwards got another letter, stating that Dr Russell was quite satisfied with what I had said. The letter contained a present, and a promise that if his influence could ever be of service to me he would be happy to use it."

This version differs from that told by Dr Russell as to his escape and recovery; but at the time he wrote his famous

"Diary" he had heard nothing of the statement made by Robb. It was a startling experience of campaigning, and shows that the exposure to danger of the war correspondent did not commence either at Tamaai or Abu Kru. The attack of the Sowars was speedily repulsed, the Cavalry completing their rout.

Next day Sir Colin delivered his attack on Bareilly in force, and the place was practically cleared of the enemy. They hovered about for days, however, and on the 7th of May it was found necessary to dislodge some who had either remained in the city or returned to it. For this purpose a party of the 93rd, under Brevet Lieutenant-Colonel Ross, was employed. The fight was a stiff one, and Sergeant Fiddes had a narrow escape. He was engaged in close combat with several of the enemy—one of whom had run his bayonet through his shoulder—when Lieutenant R. A. Cooper, of Secundra-Bagh fame, made his appearance, and rushed to the sergeant's assistance. With his revolver he shot three of the rebels dead on the spot. The rest were quickly disposed of.

The pitched battles in which the Highland regiments had to engage in this campaign were now for the most part over. Their courage and endurance were for months after severely tested by heavy marches through "devil's breath," sand storms, harassing, desultory fighting, the effects of climate, and inroads of disease; but the rebels were now demoralised, and showed little disposition to meet the onslaught of a British Force. One act of conspicuous heroism, however, yet remains to be told to the credit of the 42nd.

Chapter XXI.

THE incident to which we referred at the close of the last chapter took place on the 15th of January 1859. By that time the rebellion was well-nigh crushed, and the work of the British columns was chiefly to restore order and peace. A small body of troops was sent to Nepaul to beat into submission or scatter some insurgents who were still causing the authorities trouble. The enemy having been pursued into the Khyreegurh district, attempted to force their way into Rohilcund. On the morning of the 15th January they crossed the Sarda river about three miles from the camp of Colonel Smyth, of the Bengal Artillery, who had under him, along with other troops, Captain Lawson and No. 6 Company of the 42nd Highlanders. As soon as the alarm was given, the troops in Smyth's camp moved out with all speed, and in the jungle attacked the rebels—2,000 in number, and armed with two small field guns. Ensign Coleridge, in command of 40 men of Lawson's Company, and some untried Natives forming part of the Kumoan levy, were so placed as to be cut off from the remainder of the Force, and so did not share in the active work of the engagement until late in the afternoon. The burden of the battle was thus thrown upon Captain Lawson and the 37 gallant Highlanders under his command ; for, unfortunately, the Native Cavalry and levies who accompanied Lawson either did not fight at all, or threw in their

fortunes with the rebels.   The battle was fierce and
determined.   The mass of the enemy again and again tried
to break through the thin line of Highland skirmishers, but
each time they were repulsed with heavy loss.   At the
opening of the action reinforcements had been sent for, but
hours passed by, during which the battle raged, and no aid
came to the sorely beset Highlanders.   Yet they stood their
ground with unflinching determination.   Lawson, one of the
bravest of the brave, who had joined the regiment as a
private, and by sheer force of courage and character had
gained his present position, showed an example of the most
splendid heroism to his men—by whom he was greatly
beloved.   He directed their movements in the very forefront
of the battle.   When he despaired of the arrival of the
reinforcements, he turned to his men and said—" Lads!
we'll have to do this ourselves."   He was answered with a
ringing cheer.   More than once the Highlanders assumed the
offensive—Lawson taking command of the centre of the
company, and Colour-Sergeant Landles the left—and charged
the enemy lurking in the thick jungle—bayonet fighting with
tulwar.   As the day wore on the men got terribly exhausted
—it was incessant fighting, without a rest, without a mouth-
ful of water or a bite of food; and every now and again
a poor fellow sank to rise no more.   Says Private Robb,
whom we have already quoted, and who was present at the
engagement—" I was present at the Alma, and had seen
some hard fighting in India ; but I never gave myself up for
lost till that day in the jungle.   Our ammunition had more
than once given out, and when it was mentioned to Lawson
that it was getting short again, he shook his head and said
' That's true.'   At length he fell with his death wound
in the groin.   Pay-Sergeant Landles fell, too, and the fiends
did not stop till they had hacked him to pieces, as butchers

would cut up meat. They did the same with four others of our comrades. We had seven Irish lads in our company that day, and well indeed did they uphold the character of the corps. One of them, James Bougler, who is still alive, fought like a lion himself, and kept calling out to the rest of us—'Now, my lads, let's sell our lives as dearly as we can.' Dr Smythe, the medical officer in charge of the party, not only attended to the wounded, but got hold of a rifle and fought like a brick. A plucky little drummer-boy also did us good service by carrying ammunition." When Lawson and Landles had fallen, for some time the command devolved on a lance-corporal; but Sergeant Barclay subsequently took command, and the men were formed in line and remained on the defensive, the old soldiers cheering and keeping up the spirits of the young ones by exhortation and example. The fight went on without abatement of fury till the shades of night were drawing in, when the welcome sound of reinforcements was heard in the jungle behind—and, turning, the sadly beleaguered survivors of No. 6 Company saw No. 7 advancing to their aid. They were in skirmishing order, and, says Robb, his admiration probably excited by the pressure of circumstances—"I never saw a company advancing so beautifully." This timely succour put an end to the conflict. The enemy, baffled, re-crossed the stream in the dark, leaving many dead and dying on the ground so stubbornly disputed by the handful of Royal 42nd men. They also left their two small guns and some cattle on the field from which they had been driven. The Victoria Cross was given to Privates Duncan Miller and Walter Cook for the exhibition of conspicuous bravery during the action. Said General Walpole in his report :—"The conduct of Privates Cook and Miller deserves to be particularly pointed out. At the time the fighting was severest, and the few men of the 42nd Regiment were so close to the enemy, who were in great numbers, that some of the men were wounded by sword cuts, and the only officer with the 42nd was carried to the rear severely wounded, and the Colour-Sergeant was killed, these soldiers went to the front, took a prominent part in directing the Company, and displayed a courage, coolness, and discipline which was the admiration of all who witnessed it." For their conduct on this occasion Sir Colin paid the survivors a high compliment, speaking of them as a pattern of valour and discipline. Sir Hugh Rose, who succeeded Sir Colin as Commander-in-Chief of the Forces in India, in a stirring address, said the conduct of No. 6 Company proved "that the spirit of the Black Watch of 1729 was the same in 1859." Two days after Sir Hugh called the men of the company to the front, and personally thanked them for their gallant conduct, and each man had it recorded in his book that he was present at the engagement. Five survivors of the gallant band—Charles Sanderson, Robert Sibbalds, James Brown, James Logan, and Alexander Robb— are at present, we believe, residing in Dundee. *

One cannot help wishing that poor Lawson had lived to receive the honour his heroic conduct merited. But his death was the death of a brave soldier's ambition—in the forefront of the battle, with his gallant lads fighting by his side. This action places in a clear light the invincible valour of British soldiers. If ever excuse might be found for men giving up the contest in face of overwhelming numbers, this was such a time. But not a soldier in all that brave band wavered. When all with authority to lead had fallen, each man felt greater and nobler as he realised his personal responsibility in maintaining the conflict and defeating the foe. The discipline was as perfect and the valour as great as could have animated the Spartans who died at Thermopylæ.

Q * See Appendix G.

We can follow but a very little further the fortunes of the Highland Regiments in this trying campaign. At a late stage the 72nd and the 74th arrived on the scene, and performed some very useful work, including not a little hard fighting. But the burden of the struggle was by that time lighter, and it was not the fortune of the gallant regiments named to bear so heavy a share as they could have wished. Still they bravely sustained the fatigues of long rapid marches, and preserved their discipline in a manner that called forth the highest encomiums of their leaders. Wherever they fought they were victorious, and not only victorious, but wrested victory from the enemy when less determined troops would have failed. Lieutenant Vesey's march of 3,000 miles, with his flying detachment of the 72nd, was a signal triumph of pluck and endurance.

The 42nd remained in India until January 1868, when it sailed for home, accompanied by Colonel Priestley. Portsmouth was reached on March 4th, when five companies were immediately despatched to Perth and four to Dundee. In the succeeding October the regiment was moved to Edinburgh, and was received with acclamation by the citizens. Colonel Priestley, a good officer, much beloved by the men, was by this time dead, and Brevet Lieut.-Col. M'Leod was in command, assisted by Majors D. Macpherson and F. C. Scott. Such a turnout of spectators had not been seen in the streets since the welcome given to the 78th when they had arrived from the same land—fresh from the campaign—nearly ten years before. On the men emerging from the station, the band of the regiment struck up "Scotland Yet," and cheer on cheer rose from the assembled thousands. Along the route to the Castle the streets and every point from which a view could be obtained were lined with people proud of their "returning brave." The Castle Esplanade and even the battlements and embrasures of the Castle itself were literally filled, and as the head of the regiment approached, the sound of the martial music was drowned in the tumultuous cheering which burst forth, again and again renewed.

On April 2nd, 1872, the magnificent mural monument, which is the subject of our frontispiece, to the memory of the fallen of the Black Watch was unveiled in Dunkeld Cathedral. The work was designed by Sir John Steell, and was subscribed for by the officers of the Regiment. It is executed in white marble, and the figures are life size. The central one represents an officer standing on the field with uncovered head, gazing on the body of a heroic young ensign, who lies dead at his feet, his hand still clasping the staff of the flag he had died to save. Underneath, besides a record of the chief battle-grounds of the regiment, is the following inscription :—

IN MEMORY OF
THE OFFICERS, NON-COMMISSIONED OFFICERS,
AND PRIVATE SOLDIERS
OF THE
42ND ROYAL HIGHLANDERS—THE BLACK WATCH—
WHO FELL IN WAR
FROM THE CREATION OF THE REGIMENT
TO
THE CLOSE OF THE INDIAN MUTINY,
1859.

The Ten Independent Companies of the Freacadan
Dubh, or Black Watch, were formed into a
Regiment on the 25th October 1739, and the
First Muster took place in May 1740,
In a field between Tay Bridge
and Aberfeldy.

Here 'mong the hills that nursed each hardy Gael,
Our votive marble tells the soldier's tale ;
Art's magic power each perished friend recalls,
And heroes haunt these old Cathedral walls.

The ceremony of unveiling the monument has been well described as " one of the most interesting events in connection with the history of the Black Watch." It was the inauguration of the first public memorial of a long record of glorious deeds which has, we believe, no parallel in military history. A detachment of the 42nd, under the command of Major Duncan Macpherson, younger of Cluny,* came from Devonport to perform the ceremony of handing over the monument to the custody of the Duke of Athole, and to place over it the colours which the regiment had carried throughout the Mutiny. In the vestibule of the Cathedral a large and distinguished company had gathered, and upon entering Major Macpherson placed the old colours of the regiment over the monument. When the monument had been unveiled by the Duchess-Dowager of Athole, Major Macpherson said :—

" May it please your Grace, ladies and gentlemen, we, a detachment of the 42nd Royal Highlanders, have come here to deposit the old colours of the regiment in Dunkeld Cathedral—a place which has been selected by the regiment as the most fitting receptacle for the colours of the 42nd—a regiment which has been essentially connected with Perthshire. In the name of the officers of the regiment, I have to express to his Grace the Duke of Athole our kindest thanks for the great interest he has taken in this memorial, which I have had the great honour to ask the Dowager-Duchess to unveil ; and if I may be allowed, I would express to your Grace the kindest thanks of the regiment for the great interest the late Duke of Athole took in this monument."

In his reply, the Duke of Athole said that Major Macpherson and the detachment of the 42nd had paid a great compliment to the county of Perth, and the district of Dunkeld in particular, by placing this beautiful memorial and

*See Appendix H.

those battle-worn colours within the Cathedral. He assured them that the utmost would be done to preserve the monument from all harm. His Grace remarked that one of the earliest colonels of the regiment—Lord John Murray—was one of his own family; and at different times many men from Athole had served in its ranks. Many relations and friends of his own had likewise served with the regiment. He asked Major Macpherson to convey the thanks of the county of Perth to the officers of the 42nd for the honour they had done to the county.

A lunch and further speech-making took place subsequently, at the residence of the Dowager-Duchess, and the celebration was most enthusiastically conducted to a conclusion.

As this volume is going through the press, a movement is being carried through to erect yet another memorial to this gallant regiment, to commemorate its muster in 1740, and to mark the spot, about which there is some dispute, on which the muster took place.

Chapter XXII.

IT was not till March 1870 that the 93rd returned to Scotland, after the harassing duties of the Indian Mutiny. From the suppression of the mutiny until their return the regiment had undergone much hard work, and experienced many vicissitudes. Ten years before, Sir Colin Campbell, returning home after achieving his greatest professional triumph, had received a farewell letter from the non-commissioned officers and men of the regiment, to which he replied that it was a pride to him that his own regiment, with which he had been so long associated in scenes of war, and in hardships and fatigues innumerable, should not forget him at the last. He assured them he should not forget them while he continued to live. He was laying away his sword from a weary hand; but they were young, strong, and fit for duty. And he should watch them with an attentive eye, and should always feel for their name and glory as if they were his own. And now Sir Colin (Lord Clyde) was in his last resting-place, having carried the field-marshal's baton—the highest military rank—ere the last chapter of his devoted life was closed.

The regiment had suffered severely from disease—cholera having raged fiercely in its ranks while lying at Peshawur during the summer and autumn of 1862. Officers as well as men were attacked, and one gallant and able soldier—Major Middleton—succumbed, causing a "great loss, not only to the 93rd, but to the service at large." Middleton's death was followed by that of Lieutenant-Colonel Macdonald, Ensign Drysdale, and Dr Hope. At this time the rank and file were dying off like flies; but Major Burroughs—a tried soldier, well trusted by the men—having succeeded to the command, marched the regiment across a low range of hills to a place called Jubba, where the epidemic quickly abated and finally disappeared.

During this time the regiment had borne up with the utmost fortitude. Says Captain Burgoyne—one of those who went through the plague—"There was everything to depress the men. They had seen comrade after comrade taken by the ' pestilence which walked in darkness' among them. The great majority of them were greatly weakened by long-continued fevers. Scarcely a man but felt the workings of the cholera poison in his system, its presence being indicated by constant nausea, giddiness, difficulty of breathing, and cramps in the legs or arms. Notwithstanding this, however, there was never any approach to panic, no murmuring or shrinking from duties the most trying or irksome. At one time the same soldiers would be on hospital fatigue almost every day, rubbing the cramped limbs of groaning and dying men. Yet they never complained, never held back in a single instance, so far as is known. So long as their strength held out, they not only performed the duties assigned them willingly, but with a kindness, tenderness, and devotion which can never be forgotten by those who witnessed it."

Their staff doctors, headed by Dr Munro, worked day and night through this awful period, exhibiting great zeal and resource. When at last the epidemic lifted its cloud, the survivors, as an expression of their deliverance, raised a subscription of 674 rupees, and devoted it to charity—portions

being given to distressed operatives in Lancashire, Glasgow
cotton spinners, mission purposes in India, and to the orphans
of the regiment. To the disease 93 persons succumbed out of
268 cases. The officers, men, women, and children numbered
1149, and so harassed were they with disease during this
terrible year, that during its run no less than 2665 cases were
on the doctors' sick list.

In 1863 the regiment had engaged under Sir Neville
Chamberlain in the Eusofzai campaign, which involved more
hard marching than fighting; in 1864 the promotion of
Colonel Stisted had led to the further promotion of Major
Burroughs to the Lieutenant-Colonelcy of the 93rd, and
when at length the troopship Himalaya steered into Leith,
carrying back the bronzed veterans to their native shores,
this gallant officer held the proud position of Lieutenant-
Colonel in command.

At Aberdeen, to which they were carried by train, a
splendid reception awaited the gallant fellows. Right
heartily did the people of Bon-Accord welcome the brave
Sutherland men to the Granite City. Crowds filled the rail-
way station; the streets, all the way to the barracks, were
lined with old and young, stern and fair. From every
window bright glances bade the soldiers welcome, and
garlands and flags spoke gaily, yet silently, of the people's
admiration for the heroes of the kilt. The progress of the
men was slow, for there were many welcomes and hand-
shakings — heartfelt joy at the regiment's return being
mingled with much heartfelt sorrow that so many were left
behind who would never more gaze on Scotland's purple hills.

In the following year the headquarters of the regiment
were removed to Edinburgh, and the citizens of the old
capital turned out in their thousands to bid their heroes
welcome. Here new colours were presented to the regiment,

FLAGS CARRIED BY THE 93RD HIGHLANDERS THROUGH THE INDIAN
MUTINY WAR.
(Now in Dunrobin Castle.)

the ceremony being performed before an immense assemblage
of people in the Queen's Park by Her Grace the Duchess of
Sutherland. Before the new flags were presented, the worn
and tattered colours which the regiment had carried throughout the Indian campaign were "trooped," Captains E. H. D.
Macpherson and Gordon Alexander carrying the proud
emblems of the regimental honour.

Her Grace performed the task with rare tact and ability,
the charm of her manner and the vigour of her language
captivating all who saw and heard. She expressed her
pleasure at being called on to present new colours to the
regiment. She now belonged to Sutherland, and loved it
and its people ; but she did not forget her own brave
ancestors who were ranged on the opposite side to Sutherland
in council and in fight. But now they were united by the
common tie of devotion and loyalty to the same sovereign,
and she prayed from her heart for a blessing on the colours
of the Sutherland Regiment.

Colonel Burroughs' reply was spoken in the firm language
of a soldier. After thanking Her Grace for the honour she
had done the regiment, and mentioning that the colours now
given up were presented by the Duke of Cambridge after the
Crimean War, he continued—" Those colours that are now
so war-worn and tattered were our rallying point in the
Indian Mutiny war. We offer them for your Grace's acceptance, and hope you will accord to them an asylum at Dunrobin Castle, where the regiment was first mustered. On
former occasions of presentations of colours, it is recorded
that the officers then in command promised and vowed, in
the name of the regiment, that it would do its duty to its
King, Queen, and country. The pages of history are
witnesses how faithfully those vows have been kept. In
accepting these new colours at your Grace's hands, I call

upon the officers, non-commissioned officers, and soldiers of the 93rd Sutherland Highlanders to bear in mind that they were presented by Her Grace the Duchess of Sutherland, and I call upon the regiment to vow with me that we will defend them to the last; that we will ever faithfully do our duty to our Queen and country; that we will never permit the good name of the Sutherland Regiment to be sullied, and remembering that the Sutherland motto is *sans peur*, that it will ever be our endeavour that our conduct on all occasions shall be *sans reproche*."

Her Grace returned the Queen's colour to the regiment that it might be placed over the memorial in St Giles's Cathedral to the officers and men who fell in the Crimea; but it was afterwards decided by the vote of the officers of the regiment —25 for and 11 against—that the flag should be sent to Dunrobin, to remain side by side with the old regimental colour. It was subsequently conveyed to Dunrobin in charge of Quartermaster Harry Macleod, a Sutherlandshire man, who had fought under it "in all the battles and campaigns in which it had waved over the regiment."

The flag was received with much ceremony at Dunrobin— Macleod being escorted to the castle by the local Volunteers —and it now rests side by side in the castle with the old regimental colour.

We have now ended our narrative of the conduct of the Highland Regiments engaged in the stress and struggle of two great campaigns. It is very imperfect, very far short of setting in the best light the heroic conduct of the brave men who exposed their lives so freely for the cause of their Queen and the honour of their country. But so far as it has gone it has been an unsullied record of heroism—the story of men who never lost sight of the paramount demands of honour and duty. Once more in these terrible struggles for

supremacy the men who fought under the Highland flags
showed they had lost nothing of the conspicuous daring and
valour which distinguished the Highlanders at Fontenoy,
Aboukir, Quatre Bras, and Assaye.

They had been—what Highland soldiers ever must be—led
by able and brave officers, who never refused to share the
dangers they called on the rank and file to endure.   It is this
quality which gains the Highland soldier's heart, and makes
him the obedient slave of the man he follows.   It was this
feature of Sir Colin Campbell's leadership that endeared him
to all ranks—and notably the Highlanders.   With them he
could lie down on the bare ground, his cloak for his covering,
his saddle for his pillow.   With them he could go into battle
unbreakfasted, and with them brave the storm of the enemy's
fire.   And when the severe tension of duty was for the day
or the hour relaxed, he could pass a joke, not only with the
aspiring subaltern, but with the humblest in the ranks.   On
the march he was everywhere, inspiring the strong, cheering
the wayworn and the lagging, and speaking words of
sympathy to those who were compelled to sink under the
weight of duty.   He imbued others with the same feelings as
those animating himself; hence the confidence which subsisted
between officers and men, and the consequent triumph of
every assault made upon the enemy, who on all occasions far
outnumbered the British strength.   Our narrative has shown
with what daring all acted while the time of trial endured,
and how men, both in and above the ranks, again and again
gained the tempting and cherished reward of conspicuous
heroism.

We shall now, in the succeeding chapters, very briefly describe the more recent services of the Highland Regiments.

Chapter XXIII.

THE 42ND IN ASHANTEE.

HE task of giving Coffee Calcalli, King of Ashantee, a better idea than he possessed of the strength of Great Britain was a light one compared with the mighty effort of quelling the Indian Mutiny; yet it was in all conscience severe enough to the slender forces engaged. In 1873 Ashantee was groaning under perhaps the most brutal and bloodthirsty despotism that ever left its horrid traces on the face of the earth. In Coomassie, the capital, there were no rejoicings like the wild revelries of bloodshed—no aromas so grateful to the nostrils as the foul odour of the putrefying dead. Bloodshed was the people's chief delight—men, women, and children being often slain in pure sportive wantonness. The floors and walls of the King's palace were coated with human blood—the life-blood of thousands of victims who perished year by year that the King might find enjoyment, and the populace be furnished with interesting spectacles. In the streets headless corpses lay strewn about, swelling and bleaching; but these were of no interest since their blood had ceased to flow, for it was the spouting forth of the red life-stream that was the chief element of joy to this ruthless people. Animal and vegetable life decayed and died in the pestilential atmosphere ; but the people flourished and waxed fat in this " metropolis of murder."

But King Coffee wanted a new excitement, and in January of the year named he marched with 60,000 of his warriors— brave and warlike men—upon the British station at Cape Coast Castle. The force at command of the Governor there was inadequate to the task of giving King Coffee the drubbing he deserved, and it was decided to send out a British Force under the command of one who was well fitted to the task, although at that time the youngest General in the army, Sir Garnet Wolseley. The object of the expedition was to relieve the pressure upon Cape Coast Castle, to carry the war into the Ashantee territory, and strike a decisive blow at Coomassie, the seat of Ashantee power, which at this time had a populace of between 20,000 and 30,000, and whose King was a menace and a terror to the tribes for hundreds of miles around.

The Black Watch, commanded by Sir John Macleod, was one of the line regiments sent after the Commander-in-Chief, its strength being considerably recruited by volunteers from the gallant Highland 79th. The nature of the work to be done by the expedition will be to some extent understood by the following lines from the preface to Mr H. M. Stanley's "Coomassie and Magdala" :—"Coomassie was a town insulated by a deadly swamp. A thick jungly forest—so dense that the sun seldom pierced the foliage ; so sickly that the strongest fell victims to the malarias it cherished—surrounded it to a depth of 140 miles seaward, many hundred miles east, as many more west, and 100 miles north. Through this forest and swamp, unrelieved by any novelty or a single pretty landscape, the British army had to march 140 miles, leaving numbers behind sick of fever and dysentery."

Sir Garnet chose Cape Coast Castle as his point of departure, and immediately commenced operations by sending forward provisions to the front, and stockading posts along the main road to the Prah river. By the aid of Natives, chiefly Fantees, sworn foes of the Ashantees, and a number of Houssas, most of the work had been accomplished, when, in the early days of January 1874, the 42nd

General Lord WOLSELEY, G.C.B., K.C.M.G., &c.

landed at Cape Coast Castle, in readiness to go forward into the deadly jungle. They had left behind them their picturesque kilted dress, and were, with the exception of the regimental pipers, soberly attired in Norfolk grey. Sir John Macleod being selected for a Brigade command, Majors Macpherson and Baird had command of the regiment, and no two men could have inspired the Highlanders with

greater confidence. They had campaigned together in by
gone days, and well understood each other's calibre. Among
the other officers present with Sir Garnet were men whose
names to-day are household words—Baker Russell, Evelyn
Wood, Archibald Alison, Buller, Brackenbury, Lanyon,
Hewett, Greaves, &c.

Sir Garnet himself was a man to infuse the necessary
spirit into his troops. Cool, alert, daring, calculating, yet
full of enthusiastic dash, he was one whose will moulded
other men's minds to his own. He never hesitated, never
wavered, but formulated his plans with the decision of a
man deeply alive to the responsibilities of his office, and
fully informed as to the requirements of the work in hand.

As the troops proceeded to the front, the assistance of
immense numbers of Natives was taken advantage of. They
acted as carriers and engineers' assistants, but were continually deserting in large numbers, and causing great annoyance
to Sir Garnet, and delay in the operations. *En route*, every
soldier had chocolate in the morning and tea at night, with
occasionally a little rum. Quinine was also administered
daily to prevent illness from the climate. Early in January
the expeditionary force had crossed the Prah river—one-half
of the distance from Cape Coast Castle to the capital of
Ashantee having been accomplished. The enemy were
retiring before the British, and a delay of many days took
place, during which *pourparlers* went on between Sir Garnet
and King Coffee, through the medium of Ambassadors sent
to the British camp by the latter. These came to nothing,
and the final advance commenced. Some thousands of
natives, Akims and Wassaws, who had undertaken to
operate on the flanks during the march, failed to fulfil their
promise, and the little British Force entered the dense dark
jungle to penetrate the 80 or 90 miles yet before it, practi-

cally relying on no strength but its own. Lord Gifford led
the advance-guard far in front, and gained the crest of the
Adansi Hills without meeting anything more formidable than
a priest and a few followers, who attempted to drive back
the white men by means of curses and maledictions.
Failing in this, they turned and fled into the fastnesses of
the jungle.

On January 30th, Gifford found himself a short distance
from the town of Amoaful, with the enemy in strong force in
front. The position was as carefully reconnoitred as the
nature of the ground would permit, and a halt was made to
allow the main body to come up. Quarman village was the
place chosen for the halting ground, and at seven o'clock the
next morning the head of the main column had arrived at
the village. The 42nd was the leading regiment, and Mr
Stanley, who had been at the rendezvous throughout the
night, says :—"We had but barely finished our breakfasts
and buckled our belts on, when our servants informed us
that the white troops were close by. Hastening to the
square or plaza of the village, we were in time to witness
the famous 'Black Watch' come up, all primed and ready for
action. This was our first view of the fighting 42nd Highlanders, and I must say I improved the occasion to get a
good look at them, as though I had never seen a British
regiment in my life. Their march past was done with an
earnest determined stride that promised well for their behaviour, whatever might lie at the front."

The first column was commanded by Sir Archibald Alison,
the left by Col. Sir John M'Leod, and the right by Col.
Evelyn Wood. The 42nd was in the first column, and the
chief command of the regiment devolved on Major Macpherson, yr. of Cluny. Sir Garnet himself, bright-eyed, confident, and smiling, was mounted in a Madeira cane chair,

R

carried on the shoulders of four burly semi-nude Fantees, and closely surrounded by his staff.

Around the village of Eginassie was a forest of the very densest kind. The vegetation was of so solid a nature that thousands of men might have lain concealed within a few yards and been absolutely invisible. The brush grew to a vast height, and was here and there intersected by narrow lanes, which, running hither and thither, presented a bewildering maze, in which one might wander for hours between solid walls of foliage without knowing that he had mistaken his path. In the battle which ensued, the British troops occupied the lanes—clear marks for the enemy—while the Ashantees fired from their concealment in the bush. The village of Eginassie was taken by a rush, and then just beyond the real battle commenced. The 42nd were advancing in skirmishing order, in the manner laid down by Sir Garnet in his instructions, when suddenly from left, right, and front a terrific fire was opened from the bush. Ere long the firing became tremendous—one rumbling, hoarse, hissing roar, which far and near awoke the echoes of the erewhile silent forest. Right bravely the Black Watch men held their own, though assailed on every side by a literal hail of slugs fired into them at but a few yards distance by thousands of unseen foes. Forward they went, although men were dropping at every step; forward, although the blood was already flowing from many in the ranks. Baird was severely, as it proved fatally, wounded; Macpherson was struck in several places, one slug having gone through his leg, but he supported himself for a time on a stick, and pluckily continued to go on with his men. At one moment a portion of the bush would be as silent as the grave, and apparently as destitute of life; the next it had become a blinding sheet of flame, the wild yells of the lurking Ashantees mingling with the hoarse noise of their weapons. It was trying work for inexperienced soldiers engaging in a death struggle with an enemy that could neither be seen nor numbered, and whose system of fighting was so full of unpleasant methods. But the Highlanders went on—grim and steady, firing at every spot whence they could see the shot puffs appear from the bush.

By this time the 23rd, the Naval Brigade, and one of Rait's guns had come into action, and were pressing on the concealed enemy ; the big gun—there was room for no more—firing point blank at fifty yards distance from the Ashantees.

The roar of conflict had become deafening; so deafening, indeed, was the uproar in that vast, dense jungle, that the sound of the big gun was unheard above the other noises that contributed to make hideous the din of battle. Bravely the Ashantees held their own, but the 42nd were now in high spirits. They had overcome the first terrors of the bush, and were bent on driving the enemy either into a bit of clearing that they might give them the shock of a real Highland charge, or into a part of the jungle which might prove too dense even for King Coffee's dusky warriors. Between ten and eleven o'clock the Ashantees had begun to despair of the virtues of their fetishes and the power of their arms. Signs and wonders had told them—first, that the white men would never cross the Prah ; then that they would never surmount the Adansi Hills ; and, finally, that they could not live in the malarious, deadly jungle. They had made human sacrifices, and left the headless trunks in the roadway to warn back the pale-faced intruders ; they had mustered in strength with their best warriors in line of battle, and fought with that savage tenacity which had never before failed to conquer ; but the white men would not be stayed. Here they were over the Prah, beyond the Adansi Hills, grappling with them

in the very heart of the jungle, and driving them back,
fight wildly as they might. The poor Ashantees lost faith in
themselves, their fire slackened, they wavered, got huddled in
the roadway, and then ran back towards Amoaful.

This was the turning point. The advance became more
rapid. In front of the 42nd was the Brigadier, Sir Archibald Alison, who, seeing the decisive moment come, ordered,
according to the *Daily News* correspondent, the pipes to
strike up, and the charge to be sounded. In a moment the
forest was ringing with the notes of "The Campbells are
comin'," intermingled with a wild Highland cheer, as the
splendid regiment, with bayonets for the first time levelled,
went straight at the foe. This was the moment for which
the Highlanders had with stern eagerness waited, but they
were to be foiled. The savages could not stand the sight of
that terrible array of shining steel, and with accelerated
speed continued their flight—the shots from Rait's guns
overtaking them and swiftly crashing into their midst,
while the Highlanders could not reach them.

An eye-witness of the battle of Amoaful writes :—"Never
was battle fought admitting of less description. It is impossible, indeed, to give a picturesque account of an affair in
which there was nothing picturesque ; in which scarcely a
man saw an enemy from the commencement to the end of
the fight ; in which there was no manœuvring, no brilliant
charges, no general concentration of troops ; but which consisted simply of lying down, of creeping through the bush, of
gaining ground foot by foot, and of pouring a ceaseless fire
into every bush in front which might contain an invisible
foe. Nothing could have been better than Sir Garnet
Wolseley's plan of battle, or more admirably adapted for the
foe with whom he had to deal. Where he attacked us he
found himself opposed by a continuous front of men, who

kept his flank attacks at bay, while the 42nd pushed steadily and irresistibly forward. To that regiment belong, of course, the chief honours of the day ; but all did excellently well."

The sound of battle in that dense forest produced a strange effect upon those who could only listen. Wolseley, eager and alert, kept steadily receiving information and issuing orders, carried to and fro by tireless riders, who constantly came and went. The whole movements he kept firmly in hand, no sudden development of the enemy's strategy ever for a moment catching him unprepared. H. M. Stanley, riding behind the leading column with the General, in a few pithy paragraphs, describes what could be learned by those not in the forefront of the battle :—

" We still continued to move on slowly, with the alarm of war constantly rising in volume, listening, for want of something else to do—for we were imprisoned between lofty walls of vegetation—to the crackling, ripping sound of Snider rifles, varied by the louder intonations of the overloaded muskets of the Ashantees.

" Five minutes thus, and nothing disturbed any man's thoughts, unless of those with responsibilities on their minds, who had no time to think of the ideal or the æsthetic aspect of the moment. An officer of the 42nd here rode up on a white mule to communicate with Sir Garnet, who leaned eagerly forward on his chair to hear the news from the front. He was wounded, but his intelligence was to the effect that the left column had lost touch of the left wing of the 42nd Highlanders. The officer received orders to tell Colonel M'Leod, who commanded, to cut diagonally to the right through the bushes until both columns should join.

" At 9.15 A.M. our pace quickened somewhat, until suddenly light was exchanged for the gloom of the forest shade, and we entered the village of Eginassie, which the scouts of Lord Gifford had carried with a rush at 8.5 A.M.; and, oh, what a scene! We, travelling so soberly and unexcitedly through the narrow avenue leading through the forest, had not thought of this—the possibility that while we listened to the rumbling and muttering of the musketry, there were fellow-creatures being smitten to death in the dark recesses of the forest around Eginassie.

" Right at the mouth of the umbrageous pass the wounded and dying and dead sat grouped or lay stretched on the ground, attended by the kind ministering hands of their white brothers. It may be that the readers as well as the writer of this have seen many a gory battlefield, but I must confess that I felt much affected at the sight of the first blood drawn. . . .

" Proceeding up the hill, with our minds dwelling on the great tragedy of the day, we arrived at Amoaful. The town is conveniently situated on the extensive tabular summit of a low hill, the sides of which slope into thin threads of ooze and marsh and running stream, all clothed in impenetrable underbrush and dense growths of tropical plants and trees. A broad street or avenue (about 60 yards wide) running the whole length of the town, a distance of about 400 yards, divided it into equal portions.

" As I entered the town the aspect of the principal avenue was a busy one. Under one great tree, reclined on benches, or cots taken from the houses, the wounded officers of the 42nd Highlanders. Major Duncan Macpherson, who commanded the 42nd under Sir Archibald Alison, was rubbing his wounded leg, and doubtless thinking what a narrow escape he had from losing his poor limb."

The total number of British casualties amounted to about 250, to which the 42nd contributed considerably over 100 in killed and wounded—Brevet-Major Baird being one of those who succumbed to their wounds. Two Highlanders who fell close to the enemy were instantly beheaded. A writer, quoted by James Grant in his "British Battles," says—"The story of what became of one man will illustrate the desperate nature of the fighting. The companies were fighting in sections, each company being divided in four. One section of Captain Baird's Company had every man wounded except four—the Captain himself being among the number. Two of this section carried him to the rear, and two unwounded men only were left. These had got separated in a thick part of the bush. One of them came across a cluster of Ashantees, who all fired at him, and he fell covered with wounds, and his head was immediately cut off and carried away. The 42nd pushed on, and some of Russell's Regiment shortly afterwards came across the body, and it was sent into Amoaful for interment."

Another unfortunate Highlander on the extreme left flank of his regiment, finding himself too severely wounded to go on with his comrades, turned with the view of retracing his steps to the village of Eginassie, where he hoped to get his wound dressed. He lost his way apparently, and fell in with a party of the enemy. They overpowered the poor fellow, and, in accordance with their usual custom, cut his head off. But the brave lad had made a determined struggle for his life, a fact which was well evinced by his slashed hands and almost severed fingers.

In his report, Sir Garnet said :—" Nothing could have exceeded the admirable conduct of the 42nd Highlanders, on whom fell the hardest share of the work. As Colonel M'Leod was in command of the left column, this regiment was led by Major Macpherson, who was twice wounded."

A night's uncertain rest at Amoaful, and on the following day the village of Becquah was captured and destroyed.

Next day the Force was six miles beyond Amoaful, every inch of the ground having been disputed ; and now Sir Garnet inclined to his left towards the river Ordah, the enemy all the while offering opposition to the advance. The Ordah was reached on the 3rd of February, and here it was found that King Coffee had determined to dispute the passage of the stream by a stubborn resistance. The King had in vain sent messages to the British General, imploring him to halt and go back, and offering to pay indemnities ; but Sir Garnet could place no reliance on the Royal word, and demanded the money in his hand and hostages in his camp as pledges of good faith. These were not forthcoming, and Sir Garnet proceeded to battle, as a preliminary to carrying out his threat of reducing to ashes the Ashantee capital.

The battle opened on the 4th of February, after a night's miserable bivouac by the banks of the Ordah in a drenching rain. The morning was dark and lowering, and daylight had a long hard struggle to dispel the darkness of the black-looking rain clouds that hung close to the earth. By seven o'clock the Engineers' bridge over the river was complete, and the Rifle Brigade, which was to have the honour of first trying the mettle of the enemy, immediately marched across. Soon the advanced portion found itself in hard conflict with the Ashantees, who had gathered into the neighbouring villages in thousands. For hours the battle raged with great fury. At nine o'clock the village of Ordahsu was captured and held, but the enemy attacked the troops there posted with the greatest fury. No progress could be made forward ; but neither would the British yield one inch of ground. Here young Lieut. Eyre, son of Sir William Eyre, fell fighting with the Native auxiliaries, murmuring the word "mother" with his last breath. Lieut. Wauchope was at the same time wounded.

Finally the enemy showed signs of wavering. They could not stand the fierce fire of the breechloaders. For seven hours by this time the conflict had raged, and the 42nd had

General Sir JOHN M'LEOD, K.C.B.

(Commanding the 42nd Highlanders in the Ashantee War.)

had little or no part of the action ; but, says Mr Stanley— " Sir Garnet now did that which he ought to have done before. He ordered up the 42nd Highlanders, and gave orders to Colonel M'Leod to carry the positions in front, and march straight into Coomassie. No man," continues Stanley, " is more cool than Colonel M'Leod in action. He drew up his men in double file from one end of the village to the other. The famous ' Black Watch' appeared, though greatly

reduced in numbers (340 on this day), to be fit followers of
their Colonel. Both Colonel and men mutually understood
each other. There was no doubt or hesitancy in either commanding officer or men. During the brief halt, Colonel
M'Leod surveyed his men, and then said—

"'The 42nd will fire volleys by companies according to
order. Forward!'

"Then began the sublime march to Coomassie, the most
gallant conduct and most impressive action of the Ashantee
campaign. The Highlanders marched out of the village into
the gloomy chasm of the forest by a road beset with ambuscades with a proud military bearing full of determination and
joyous courage."

They had gone but a little way into the dark, gaping pass,
through the dank, deadly forest, when, to right and left, the
underwood became suddenly alive with flame, and the hail
of Ashantee slugs and bullets smote the hurriedly-advancing
Celts. Now clear and loud came M'Leod's voice above the
rattle of the musketry—

"Company A, front rank, fire to the right; rear rank, fire
to the left. Forward!"

And so on the order ran along the companies; then the
sound of the fire from the Highlanders' rifles mingled
and rang in response to the fire of the enemy. Simultaneously the men burst into wild cheers as the Scottish bagpipes joined in the medley of uproar which was making the
forest ring and echo. Volley after volley crashed through
the jungle into the ambuscades; and as they delivered
their fire the brawny Highlanders marched on with swinging
pace, as if heedless of the lurking foe which in his thousands
swarmed around and were left behind. Sometimes clusters
of the dusky savages would appear in the path in front, then
the flashing steel would be levelled, and the 42nd would

advance at the charge, accelerating the flight of the already half-frightened Ashantees.

Such a mode of fighting the poor blacks had never seen ; never had they met a foe whom neither fetish nor sacrifice nor weapon could turn ; and they began to abandon the hope of success. They could not understand the white men firing into their ambuscades, then passing on, as if they were not worthy of another thought. The impression beset them that some great calamity was threatened from those men who had fired, marched on, and were now lost to view in the dense forest far in front. The fear spread, and soon the blowing of horns all around seemed the signal for flight.

Meanwhile the 42nd held grimly on to the capital, and as the Commander-in-Chief toiled on far in rear with the main body he received a cheering despatch from Sir Archibald Alison, which said—

"We have taken all the villages but the last before entering Coomassie. The enemy is flying panic-stricken before us. Support me with half the Rifles, and I enter Coomassie to-night."

The performances of the 42nd on the day of the capture of Coomassie were magnificent. They went on with the air of men against whom resistance was useless—under the most trying circumstances displaying a courage and maintaining a discipline that would have sustained them against the best troops in the world. In one of his letters, written to the *New York Herald* during the campaign, Mr Stanley gives a graphic description of the advance through the jungle :—

"Sir Archibald Alison was the Brigadier commanding the advance, but Colonel M'Leod was the officer in immediate command of the 42nd. Sir Archibald was more of a looker-on upon the exciting scene of the advance. The conduct of the 42nd Highlanders on many fields has

been considerably belauded, but mere laudation is not enough
for the gallantry which has distinguished this regiment when
in action. Its bearing has been beyond praise as a model
regiment, exceedingly well disciplined, and, individually,
nothing could surpass the standing and gallantry which
distinguished each member of the 42nd, or the Black
Watch. They proceeded along the well-ambushed road as if
on parade, by twos. Vomiting out two score of bullets to
the right and two score to the left, the Companies volleyed
and thundered as they marched past the ambuscades,
cheers rising from the throats of the lusty Scots until the
forest rang again with the discordant medley of musketry,
bagpipe music, and vocal sounds.   Rait's Artillery now
and then gave tongue with the usual deep roar and crash, but
it was the audacious spirit and true military bearing upon the
part of the Highlanders, as they moved down the road
towards Coomassie, which challenged admiration this day.
Very many were borne back frightfully disfigured and
seriously wounded, but the regiment never halted nor
wavered ; on it went, until the Ashantees, perceiving it
useless to fight against men who would advance heedless
of ambuscades, rose from their coverts and fled panic-stricken
towards Coomassie, being perforated by balls whenever they
showed themselves to the hawk-eyed Scots.   Indeed, I only
wish I had enough time given me to frame in fit words the
unqualified admiration which the conduct of the 42nd
kindled in all who saw or heard of it.   One man exhibited
himself eminently brave among the brave men.   His name
was Thomas Adams.   It is said that he led the way to
Coomassie, and kept himself about ten yards ahead of
his regiment, the target for many hundred guns ; but that,
despite the annoying noise of iron and leaden slugs, the man
bounded on the road like a well-trained hound on a hot scent.

This example, together with the cool, calm commands of Colonel M'Leod, had a marvellous effect on the Highland battalion—so much so, that the conduct of all other white regiments on this day pales before that of the 42nd. Village after village along the road heard the disastrous tidings which the fugitives conveyed, and long before the Highlanders approached the place where the King had remained during the battle, the King had decamped because of these reports."

At length, as the shades of night was falling, the brave Highlanders, flushed with victory, entered Coomassie, the dread capital of the Ashantee kingdom, and quietly took possession of the town. The King and the greater number of the people had fled; those who remained looked on the whitefaced invaders with an appearance of stolid dismay. Later in the evening the Commander-in-Chief entered the town, and the Highlanders, drawn up in a line in the main street, greeted him with thundering rounds of cheers. King Coffee was crushed; the campaign was ended.

Sir Garnet, starting on the 6th of February, marched back through forest, jungle, and deadly swamp to Cape Coast Castle, which was finally reached on the 19th. All had suffered from the fearful climate, but the naturally robust constitutions of the men of the Black Watch quickly reasserted themselves, and when the men, clad now in their own picturesque uniform—kilt and plaid—landed at Portsmouth from the *Sarmatian*, on the 23rd of March, they presented a tolerably hardy and healthy appearance. They were received with unbounded enthusiasm, for the fame of their brave deeds had preceded them, and the people vied with one another in testifying to their admiration for the gallant " Forty-twa."

Among the officers whom Sir Garnet Wolseley specially mentioned in his despatches for prominent services were Colonel M'Leod, C.B., who was afterwards made a K.C.B. ; Majors Macpherson and Scott ; Captains Farquharson, V.C., Furze, and Kidston ; and Lieutenant Wauchope. Majors Macpherson and Scott were made Lieut.-Colonels and C.B.'s, and Captains Bayley, Farquharson, and Furze were made Brevet-Majors. But the chief distinction fell to the lot of Sergeant Samuel M'Gaw—a brave Ayrshire man—who was awarded the Victoria Cross for conspicuous valour at the battle of Amoaful. M'Gaw was a well-tried veteran. He had joined the regiment in India during the Mutiny, was attached to No. 6 Company, and went through the dreadful engagement in the Indian jungle at the Sarda river on the 15th January 1859.

Other soldiers, for " distinguished conduct in the field," were presented at Windsor with special medals by Her Majesty the Queen. These were—Sergeant-Instructor Wm. Street, Sergeant Henry Baxter, Private John White, Geo. Ritchie, Geo. Cameron, Wm. Bell, Henry Jones, Wm. Nichol, Thomas Adams, and Piper James Wetherspoon. For " meritorious services" during the campaign a medal was also awarded to Sergeant-Major Barclay.

Thus once more the Black Watch earned that distinction among heroes which it has ever been the aim of the gallant regiment to obtain. The vicissitudes of climate did not diminish their strength, nor the odds against them their courage. With the same invincible discipline which they had retained all through the years, they marched against the Ashantee hordes ; and when they met them sent them back reeling and beaten, as they had already sent back the soldiers of every race against whom their Sovereign had required them to wage battle.

Chapter XXIV.

THE AFGHAN CAMPAIGNS.

WE have now reached a period when events are so recent and details are so familiar to all readers that it is unnecessary to write of them at the same length as we have done of the achievements already recorded. But they cannot be passed over. Some of the most brilliant work ever performed by Highland soldiers has been witnessed during the last nine years. In Afghanistan, in the Transvaal, in Egypt and the Soudan, they have been present in crucial battle tests, and have borne themselves with even more than their traditional heroism. It has been their fortune alike to share in brilliant victories, to avert disaster, and to sell their lives as did their fathers of old, when victory was impossible, and disaster was inevitable. Only once has the gallant Highlander in recent years been placed in the circumstances last described—on ill-fated Majuba Hill—and there he died with his weapon clutched in an iron grasp, and his face turned fearless on the triumphant foe.

When Shere Ali, the Ameer of Afghanistan, in 1878 refused to permit a British Mission to proceed to Cabul, and turned back Sir Neville Chamberlain and Louis Cavagnari at the entrance of the Khyber Pass, British power and authority in the person of Lord Lytton, the Viceroy of India, felt deeply injured, and his Lordship, with a promptitude that was very refreshing to the "Jingoes" in search of a "scientific frontier," sent an ultimatum to the Ameer. As it was not answered within the specified time, British troops marched from the north-west Indian frontier into Afghanistan.

One of the columns advanced by the Koorum Valley, and included in its ranks a sprightly Highland regiment in capital condition—the 72nd, Duke of Albany's Own Highlanders— a regiment not wearing the kilt, but the tartan trews ! The 72nd, like all the other Highland corps, had an honourable and noble record. Their colours told of their work at the Cape, in Central India, and in the trenches of Sebastopol. They had long been famed for their splendid discipline, and in that respect had gained the high approbation of the Iron Duke himself, who had been struck with their behaviour when serving in the same army with them on the plains of Hindostan. They were commanded by Colonel Brownlow, an officer with a good Crimean and Indian reputation, and were ready for anything that might be required.

In command of this Force was one of the most able officers in the British army—Major-General Sir Frederick S. Roberts. The General is the son of Sir Abraham Roberts, whose name is well known as one of the band of hard-working soldiers who exerted themselves so well to consolidate British power in India during the early part of the present century. General Roberts had served his apprenticeship to war as a Lieutenant of Artillery in the Indian Mutiny ; he had been wounded at the assault on Delhi, where his horse was shot under him. He was in the forefront of the battle whenever an opportunity was given, and had another horse killed and one wounded in the course of the campaign. At the relief of Lucknow he had shown conspicuous abilities, allied to extraordinary courage. At Khoda Gunj he won the Victoria Cross, under the following circumstances :—While

the Bengal Artillery was following up the retreating enemy,
Lieutenant Roberts saw in the distance two Sepoys going
away with the standard.   He put spurs to his horse, and
overtook them just as they were about to enter a village.
They immediately turned round and presented their muskets
at him—one of them pulling his trigger.   The cap snapped,
and the gallant young officer cut down the standard-bearer
and captured the standard.   He also on the same day cut

GENERAL SIR FREDERICK ROBERTS, V.C., &c.

down a Sepoy who was engaged in combat with a Sowar,
killing him on the spot.   Roberts had gained further
experience in the frontier wars of 1863, and in Abyssinia
under Sir Robert Napier.   In 1881-2 he was once more in

s

the field as Assistant Quartermaster-General and Senior Staff Officer with the Cachar column sent to punish the Lushais. He was full Colonel when called to the rank of Major-General and the command of the Koorum column.

The march to Cabul, the Afghan capital, was slow and tedious, and some of the tribesmen were anything but friendly ; but everything went on with tolerable ease till the Peiwar Kotal was reached, when the enemy was discovered in force. The country was in many respects like our own Highlands in winter. Snow was falling ; the cold was intense ; high mountain ranges rose on every side, and through the valleys tumultuously rushed swollen, foaming rivers. All along the Natives had from high up points of vantage watched the British advance, and now that the Peiwar Kotal was reached, they could be seen in strength on the hills in front and on the flanks.

In front of the British Force " was a valley up which the road to the Kotal wound for about two miles from the camp. Across the summit or saddle of the steep ascent the enemy had thrown up a battery of field works, the fire of which could rake the whole pass. On either side of the Kotal, on two steep hills, were guns in battery which could throw a deadly cross-fire upon an ascending force. The troops of the Ameer occupied the entire line of the upper hills for a distance of four miles, and at either extremity were guns in position to meet any flank attack that could be made, and lofty and more inaccessible hills covered their line of retreat."

Carefully did General Roberts study the position, and he finally discovered where its weak points lay. Upon this knowledge he based his plans for attack. He was firmly convinced that the position could not be carried in front, and determined, under cover of the night, to perform

with a portion of his troops a hazardous flank march, and turn
the left of the enemy's position from the Spin Gawi Pass.
The 72nd Highlanders were selected as part of the force to be
employed in this work, and General Roberts accompanied in
person.

On the night of the 1st December the movement was
made. Without sound of bugle the troops fell in, and
silently marched off to ascend the heights from which the
Afghans were to be driven. The night was bitterly cold, and
they were at a high altitude, over 8,000 feet above sea level;
yet the men bore the cold without a murmur, the Highlanders keeping up their spirits by cheerily-whispered words
of encouragement to one another. But there were treacherous hearts among the Native troops accompanying the 72nd,
and as the men were toiling up the heights, keeping themselves well under the shadows caused by the moonlight, they
were startled by the reports of two rifles from the ranks of
the 29th Punjaub Rifles, evidently fired to give the Afghans
warning of their approach.

The Punjaub Regiment was immediately halted, and the
two Companies of the Highlanders and the 5th Ghoorkas
passed to the front. Just as dawn was breaking the shrill
challenge of an Afghan sentry rang out from the silent
mountain side, and immediately the Highlanders were
engaged. Bounding forward, they found themselves confronted by an abattis rising eight feet in height, from which
an Afghan picket poured out a hot fire. The Ghoorkas went
straight at the obstacle, in aid of the slender force of Highlanders endeavouring to carry it, and a fierce conflict ensued.
For a time it was hand-to-hand—bayonet and clubbed musket
finally forcing back the fierce Afghans, who would not yield
an inch of ground till driven from it. Joined by the main
body of the 72nd, who had by a rapid flank movement

pushed itself into the fighting line, the attacking party drove back the Afghans upon a second stockade. This they abandoned for a yet stronger position on an entrenched

COLOURS OF THE 72ND, DUKE OF ALBANY'S OWN HIGHLANDERS.
(Carried through the Indian Mutiny, in Afghanistan and in the Egyptian Campaign of 1882.)

knoll, but the Highlanders' blood was up now, and in a few minutes the battle was raging round the knoll.

It was not yet daylight, and the flashes of musketry on the hillsides, the rattling noise of combat, and the wild cries of those so hotly engaged, produced a feeling of great anxiety on those who could in the semi-darkness know little of what was going on. Yet the ever-advancing fire of the High-landers' line told that they were carrying everything before them. The knoll referred to above was stubbornly held, the Afghans fighting with a desperate fury, to which 40 corpses bore evidence, when they finally retired to allow the 72nd to crown the position.

"As soon as the attack was developed," says a newspaper correspondent, who was an eye-witness of the engagement, "Captain Kelso brought up his mountain guns in splendid style. They did excellent service, but he himself was shot dead whilst fighting. Lieutenant Munro, of the 72nd, was slightly wounded. After hard fighting, lasting for three hours, the enemy's left wing was rolled up, and broke in confusion. The regiments in reserve—viz., the 2nd Punjaub Infantry, the Pioneers, and four Royal Artillery guns on elephants—came up late. Meanwhile, General Roberts proceeded towards the centre of the position. The 29th Native Infantry, leading, found the enemy very strongly posted, and a most obstinate fight ensued in the thick pine woods which cross the heights, and to which the Afghans clung most obstinately. Here Major Anderson, of the Pioneers, was killed. It was some hours before the enemy was dislodged from this point, bringing up fresh troops continually, and at times assuming the offensive, and for a while the more advanced of our troops were hard pressed, until relieved by the 5th Ghoorkas and the 72nd Highlanders. Meanwhile, the 5th Punjaub Infantry advanced under fire up the heights,

rather to the left of the enemy's centre, and the 2nd Punjaub
Infantry and the Pioneers, with the Royal Artillery guns,
had joined the General. Two mountain guns, admirably
placed, now threw their shells straight into the enemy's
camp on the top of the pass from a commanding knoll,
and rendered it untenable. But still we were unable to
advance upon Peiwar itself. General Roberts then made a
second turning movement in the direction of the enemy's
line of retreat behind their centre. This movement, combined with the admirably directed fire of Major Parry's
three field guns, far below the heights, together with the
hot fire kept up by the regiments which had now captured
the woods in the heart of the enemy's position, caused the
enemy at last to retreat in haste from their extremely strong
position, leaving behind them all their guns, a quantity of
ammunition, and all their stores. The retreat soon became
a perfect rout."

In the engagement Brigadier Cobbe was severely wounded,
a bullet going through his thigh. The victory was one of
great importance, and one evidently unexpected by the
Afghans, who had vast stores of ammunition and provisions
in camp, as if to last them a long time while they held back
the advancing British Force.

Mr Cameron, of the *Standard*, wrote, after an inspection of
the battlefield, that the Afghan dead lay thick near the top
of the ravine, where the 72nd and the Ghoorkas first met
them, and that they lay scattered through the pine woods
above and on the mountain crests surrounding the Kotal,
where they were chased in splendid style by the two
regiments named.

In his official report, telegraphed from the Peiwar Kotal on
the 3rd of December, General Roberts said :—

" We surprised the enemy at daybreak, when the 72nd

Highlanders and 5th Ghoorkas drove the Afghan troops gallantly from several positions. They afterwards endeavoured to reach Peiwar Kotal, but the assault could not be delivered on that side. We then threatened the enemy's rear, and attacked Peiwar Kotal, which was occupied at four o'clock in the afternoon. The enemy, who had been reinforced by four regiments on the previous evening, offered a desperate resistance, and their Artillery was well served. Their defeat, however, was complete. We captured 18 guns and a large quantity of ammunition. Our loss is moderate, considering the numbers to which we were opposed, and the difficult nature of the country."

After advancing to the Shutar Gardan Pass—only four marches from Cabul—General Roberts was reluctantly compelled to turn back to Fort Koorum. His Force was not strong enough to make an attempt on Cabul, and he could not remain where he was and face the rigours of an Afghan winter.

The march back was not without great trials and dangers. The 13th of December, says James Grant, in his "Recent British Battles," saw the Force marching back to winter quarters "through a five miles gorge, by a rough and stony path, overlooked by many savage heights and ridges. After a time a number of Afghans were seen perched upon these ridges, watching the troops on the line of march defiling below. . . . Before the rear of the column had quitted the ravine, more country people were seen collecting on the rocks, and when Captain F. Goad, transport officer, was walking close to a part of the small baggage guard of the Albany Highlanders, a sudden volley from above was fired upon the whole. Captain Goad fell wounded, his right thigh bone being broken by a bullet, which passed through his left leg after breaking his sword and scabbard. Sergeant William

Greer, of the 72nd, with three other Highlanders, placed him under shelter of a rock, and devoted their attention to the enemy. They were only four men against a great number, under good cover too, but they could not desert a wounded officer as long as they could defend him; and by steady and careful firing, picking off their men in quick succession, they kept the foe at bay. . . . Our casualties in this affair were one man killed, two officers, eight soldiers, and three camp followers wounded. A sick Highlander, who was being carried in a dooly, fired all his ammunition, sixty-two rounds, at the enemy; and as he was a good marksman, he never fired without getting a 'fair shot.' For his courage and devotion, Sergeant William Greer was promoted to a lieutenancy in the 72nd Highlanders in April 1879."

The campaign was now practically over for the season, but startling events caused it to be renewed with greater severity and more effect in the ensuing year.

Chapter XXV.

THE treaty of Gundamuk, signed in May 1879 by Yakoub Khan, the young Ameer, was expected to put an end to fighting in Afghanistan. The foreign affairs of the Country were placed under British control, and the country was guaranteed against foreign aggression by the aid of British money, arms, and troops if necessary; and a British Embassy, with requisite suite, was permitted to take up its residence at Cabul. The Treaty was arranged on behalf of the Indian Government by Major Cavagnari, who, for his diplomatic services, was created a Knight Commander of the Bath, and was appointed to the Cabul Embassy—his Secretary being Mr Jenkyns, a youthful but brave Scot, who had already shown excellent qualities as a member of the Punjaub Civil Service.

The Embassy took up its residence in Cabul, and soon after became little heard of by the outside world.

On the 6th of September, however, the whole country was startled by the telegraphic intelligence that the military in Cabul had risen in revolt, attacked the British Residency, and that Sir Louis Cavagnari and the whole staff had been put to the sword.

General Roberts advanced immediately by the Koorum route to Cabul, having under his command, in addition to the 72nd Highlanders, who had so distinguished themselves at the Peiwar Kotal, another Highland regiment, with a reputation

for hard service and brave deeds not excelled by the famous
Black Watch itself.   This was the 92nd Highlanders, a
regiment of veterans in splendid condition for service.   On
the colours of this gallant corps were inscribed " Egmont-op-
Zee," " Mandora," " Egypt" (with the Sphinx), " Corunna,"
" Almarez," " Vittoria," " Pyrenees," " Nive," " Orthes,"
" Peninsular," and " Waterloo."   In the terrible Peninsular
war, the stories of which are still familiar in our busy cities
and remotest glens, the 92nd Highlanders bore themselves
with a valour which made their fame imperishable.   True
sons of the mountains, they fought with a determination, a
heroism, and success which made them the wonder alike of
their enemies and of their comrades-in-arms.   Since then
they had, although much abroad, not had the share they
desired of active service, and had longed for the opportunity
of displaying how strongly the old spirit still burned within
their ranks.   Now that active work was again before them,
and they found themselves once more marching to battle,
they looked forward with enthusiasm to their first encounter
with the enemy.

We cannot detail the hard fighting which followed.   At
Charasiah the enemy in immense strength was first met, and
the brave Gordon Highlanders had their share of stern
combat.   The Afghan tribesmen are hardy mountaineers,
trained from their infancy to hardy exercises and the arts of
war.   They are brave to recklessness, strong, and athletic,
and fought with a tenacity and ferocity that precluded all
idea of quarter being asked or given—fought till their eyes
glazed and their hands refused to lift a knife or pull a trigger.
In charge of an improvised post as senior officer present was
Major G. Stewart White, of the 92nd.   Moving his men
from under cover, White saw the hills to his right lined with
the enemy in many battalions.   He directed the big guns to

play upon the hills, and then went forward with his kilted heroes. Up to this time the enemy had stood firmly against the British fire, and the Highlanders felt that to drive him from his position would require an effort of no light kind. Up they went from one steep ledge to another, clambering, toiling, but ever nearing the stubborn foe, and encouraged by the conduct of White, who went on with the leading files. Suddenly the Highlanders found a large number of the enemy straight in their front, outnumbering them by nearly twenty to one. White's men were utterly exhausted by the climbing, and could hardly go forward; but that officer, seeing that immediate action was necessary, took a man's rifle from his hand, and advancing right towards the enemy shot dead their leader. As the Afghans hesitated in dismay at this daring act and its fatal result, the Highlanders raised a loud shout, and dashed forward, driving the Afghans down the hill, and crowning it themselves with a ringing cheer. For his cool, daring deed Major Stewart White was awarded the Victoria Cross, which he had most worthily won.

While the 92nd men were giving so good an account of themselves at one part of the field, at another the gallant 72nd were leading the van of a small force operating under General Baker. The position Baker was storming was led up to by rugged and precipitous paths, up which the men toiled painfully, but resolutely. When they had been engaged for a couple of hours with doubtful success, the fortunes of the day were turned in their favour by a co-operative movement of the 92nd, who, with pipes playing and colours flying, appeared rushing up the hill on the enemy's flank, and drove him from his vantage ground. In the end, after twelve hours' hard fighting, the Afghans were driven from all their posts, their guns were captured, and the battle of Charasiah was won, with a loss to the British Force that was comparatively trifling.

Roberts marched into Cabul, and executed summary vengeance on some of those who had been prime instigators of the revolt. Then came the abdication of the Ameer, and troubles thickened. The tribes gathered in from every direction, filled with fanatical hatred against the British intruders. They attacked the troops at every weak point, and for weeks the harassing warfare went on, the British holding their own against the fierce hordes by whom they were surrounded. Lieutenant Dick Cunyngham, of the 92nd, won his Victoria Cross for holding his men together under the enemy's fire, when they, almost beaten back, paused in front of overwhelming numbers. His example and encouragement were enough for the brave lads he led, and they soon proved worthy followers of so worthy a leader. In some terrible fighting engaged in by the 72nd on the day after, Captain Spens and Lieut. Gainsford, of that regiment, lost their lives, the former when engaged in a desperate attempt to check the enemy, who at the moment were carrying everything before them. Corporal George Sellar, of the 72nd, on this day performed a valorous deed by rushing in front of the enemy and engaging in hand-to-hand conflict with a huge Afghan, who sprang from the opposing ranks to meet him. Sellar was wounded in the gladiatorial struggle, but lived to obtain the coveted Victoria Cross for his gallant deed.

The pressure had now become too great for the General to hold out in Cabul. A *jehad* or *holy war* had been declared, and the hordes were getting stronger and fiercer. A retirement to the cantonment of Sherpur, outside Cabul, was immediately resolved on, and as quickly effected. And not an hour too soon, for on the same night the Afghans in full force took possession of Cabul. 7,000 men were within the cantonments, and at once set about making everything as secure as possible against an attack which it was well known was impending. On the morning of the 23rd of December the attack came. Ere yet it was daylight a sound like the roar of the sea broke upon the ears of the beleaguered. Above the confused noise could be heard the "Ya Allah," "Ya Allah" of the fanatics. They were in immense force—from 30,000 to 60,000 strong—and they made their attack with savage fury—eager, no doubt, once more, as in 1842, to annihilate a British Force which had dared to intrude itself within their forbidding fortresses. It was for the little garrison a life or death struggle; well was its portent understood, and nobly did the men rise to the occasion. Soon the heat of combat spread from those without to those within the walls of Sherpur, and the cheers of the British and wild cries of the Sikhs mingled with the Afghan howls. The battle raged fiercely as the sun lit up the snow-clad Afghan peaks with dazzling white, and tinged the ground below, where men struggled and died, with a deep suggestive red. But Roberts added aggressive strategy to defence, and soon his counter movements through the villages in the vicinity of the cantonments began to tell with staggering effect on the Afghans. By one o'clock the enemy had become so demoralised that the General knew the time to let loose his Cavalry had come. He carried the order himself to the riders, and soon the horsemen were out and away, dealing death and destruction among the confused and scattered masses, recoiling from before the gallant British defence. The slaughter was fearful, not less than 3,000 killed and wounded being the penalty paid by the Afghans for their temerity.

In the engagement the Highlanders had fought with conspicuous coolness and precision, all their admirable qualities of courage and discipline being displayed as the conflict waxed its hottest. On Christmas day Gough's column joined Roberts', the 9th Foot and the 4th Ghoorkas quartered in the Bala Hissar. Seasonable festivities now reigned in the camp, and James Grant tells that on New Year's morning a party of the officers of the 92nd went off to "first fit" the General. Hearing cries for him, the gallant soldier came forth, and laughing, said—"The 92nd have always come to the front when *I* called on *them*, so I suppose I must do the same now." And he came forth, and, although the historian quoted averreth not, doubtless regaled himself with some of the hot whisky and water with which the gallant Gordons had been making the night festive.

We pass over the rest of the fighting around Cabul, although a gallant struggle of Brigadier Herbert Macpherson, supported by a wing of the 92nd and other troops, might well be specially detailed. On this occasion, while the whole loss of the British Force amounted to only 32 wounded, before the 92nd 100 of the enemy's dead lay prone, besides many wounded with more or less severity. Among those who had particularly distinguished themselves in the fighting just described were—Lieutenant Forbes, Colour-Sergeant James Drummond, Colour-Sergeant Hector Macdonald, Sergeant Maclaren, and Corporal Mackay—all of the 92nd; and Colour-Sergeants Yule and Macdonald, and Sergeant Cox of the 72nd. Captain Cook, V.C., a Ross-shire gentleman in the 5th Ghoorkas, also fought with distinction.

Although the tribes around Cabul were conquered, the mutinous spirit was spreading in the south and west. The troops at Candahar revolted, and General Burrows, who had, as a precautionary measure, been sent to support the Wali with a small Brigade of British troops, found himself compelled to fall back on Kushk-i-Nakud. General Burrows' Force had originally numbered about 2,300 men, including 500 of the 66th Berkshire Regiment, 500 of the Bombay

Grenadiers, and 500 of the 19th Bombay Infantry; but when he reached Kushk-i-Nakud, after some hard fighting with the mutineers, his Force did not number much more than 2,000 men. And against him was now marching, at the head of a magnificently appointed army of all arms, numbering over 12,000 men, Ayoub Khan, brother of Yakoub, and pretender to the Afghan throne under the patronage of Russia. The result was disaster to British arms. It is no part of our duty to describe the calamity that befell Burrows at Maiwand. With this sad affair we are concerned only in so far as it led to renewed effort on the part of those whose brave deeds we are briefly recording. British prestige had to be restored; the disgrace of this disaster to British arms had to be wiped out. The man best fitted to the task was quickly under orders, and it is doubtful whether the heart of Sir Frederick Roberts bounded with a higher exultation when he knew he was to lead, than did the hearts of the 72nd and 92nd Highlanders when they became aware they were to follow, in the famous march now to be undertaken from Cabul to Candahar.

Chapter XXVI.

ON the 3rd of August General Roberts received his orders from the Viceroy of India to proceed to Candahar; on the 8th the famous march began—five days only having been spent in getting a column of 10,000 men of all arms in marching order. Much depended on the rapidity of the movement, and no useless encumbrance was permitted to be taken. Yet 10,000 men cannot be marched over 300 miles of an enemy's country without a good deal of impedimenta in the way of food, ammunition, transport, and general supplies. But when everything was provided for, the column marched lightly. Two Highland Regiments, the 72nd and the 92nd, with the famous 60th Rifles, were the only British troops present, and in all numbered 1,800 bayonets.

It was no ordinary task that General Roberts had been ordered to undertake. In his mental eye, looking away forward over that long stretch of country to be traversed, with its miles of sterile rocky trackless desert, its dense jungles, awesome passes, and solitary ravines, General Roberts saw but the end, with its triumph of succour for the sorely-pressed troops at Candahar. He regarded the obstacles in the way—the burning sun, the want of water, the fatigues of the march, the ever present danger of attack from an armed and hostile populace—as mere incidents on the road to success. They had to be borne, and must be overcome.

"Our work is cut out for us," he had exclaimed when he read his orders; and the work "cut out" should be performed.

Just before the order to march off was given the General said—"The march of a division of 10,000 men over 300 miles of an enemy's country is the task I have undertaken, and which I feel confident I can carry out, relying as I do on the zeal and devotion of those who are under my command. Our march will doubtless be watched with anxiety by our friends in Candahar, and by those belonging to us at home. We must, therefore, show that British soldiers can now accomplish what their forefathers did in old times, and that upon an occasion like the present we can make any sacrifices to carry out the task set before us."

Then the column marched off, the flags of the Highland regiments being given to the breeze, and their pipers playing lustily their own inspiring native airs. We need not detail the march. It proved very trying. "For days," writes James Grant, in Cassell's "British Battles," "the August sun beat fiercely down upon the weary column, and Sir Frederick Roberts was so affected by the heat that he had a sharp attack of fever, which would have placed *hors de combat* any one else less determined to achieve the great task he had in hand." Sixteen miles a day was the allotted distance to be covered; yet the eagerness of the men would have carried them over thirty had they been permitted. But Roberts knew what was at the end of the journey, and he could not permit his troops to exhaust their energies in marching. On the march the General watched the condition of his men with a careful eye. He made it his daily business to learn how many men had fallen out of each regiment. He found that the 72nd Highlanders had more casualties than either of the other two British regiments, and that the majority of the cases occurred among the young soldiers who had last joined

T

the regiment. The 92nd stood the march splendidly, the men being nearly all veterans, the privates of the regiment showing an average service of nine years. This was no work for boy soldiers, and Sir Frederick, in a speech made in London in the following year, used it as an illustration against the present short service system. All through the Highlanders presented a striking appearance—their peculiarly fine physique, chest measurement, and muscular development being easily noticeable as exceeding those of all the other corps. The garrison of Khelat-i-Ghilzai was relieved on the way, and taking the men with him, Roberts moved rapidly on Candahar, then but a few marches distant. On the 26th of August the Highlanders distinguished themselves in a severe engagement with the enemy in a rocky defile, in which the latter were defeated with serious loss; on the 31st the column arrived in front of Candahar, and found Ayoub waiting for it in his camp at Mazra. He had with him a large and well-appointed army, admirably drilled and disciplined, and armed with many Armstrong guns.

Roberts immediately ordered an acute reconnaisance, which involved some fighting, in which Herbert Macpherson, who was then Brigadier, marched his Highlanders steadily to the front. The movement of the Highlanders evidently impressed the Afghans. They were quite prepared to try conclusions with Native troops, and would more than likely have sent them to the right about; but when they saw the demeanour of the Highlanders, threatening a direct assault in the open day, they retired as the kilted heroes advanced. Macpherson, however, did get within 200 yards of a great body of the enemy, when, opening a heavy file fire, he drove them from their position with severe loss.

The day was spent in important movements, which revealed to the General all that he wanted to know. The night was

passed in sleepless eagerness ; on the morning of the eventful
1st September all were early astir.   By half-past five the
chief officers were summoned to the General's tent, and
received their instructions clearly and concisely.   He gave
them an outline scheme, told them what he wanted done,
how he had decided to carry out his objects, and left the
details to themselves.

At half-past nine all was ready for action; the General
mounted his white horse and gave the signal.   Then the roar
of battle commenced.   The 92nd was now under the
command of Brigadier Macpherson, the 72nd under Baker.
Splendidly were both led, and brilliantly did they follow.
The 92nd were ordered to drive the enemy from the village
of Gundi-Moolah-Sahibdad, and aided by the Ghoorkas they
performed their work in grand style.   The fierce Afghans
waited their coming, and met the Highlanders' bayonet charge.
For a time the conflict was terrible, the hand-to-hand struggle
being of the most severe and exciting nature.   Rush upon
rush was made on their line by the wild Ghazees, yet the
Highlanders held firm, parrying and lunging with all their
strength, and finally, by sheer force of stubborn courage,
driving the enemy from their position.

Meanwhile the 72nd had not been idle.   They were in the
2nd Brigade, advancing in support of the first, in which was
the 92nd.   They, too, became quickly engaged in the gardens
and loopholed wall enclosures.   The fighting was desperate,
the Ghazees hurling themselves like demons on the soldiers,
dashing their shields on the bayonets, and cutting and
slashing with their long knives and tulwars.   Their ferocious
aspect added to the horrors of the scene ; but the 72nd stood
shoulder to shoulder, and exerted their whole strength against
the demoniacal savages, who strove to wrest their rifles from
their hands, and even to tear their flesh with their teeth.

In this kind of fighting death came quickly to many a poor fellow. The Highlanders lost heavily in officers. Lieutenant-Colonel Brownlow, C.B., of the 72nd, was killed leading his

LIEUTENANT-COLONEL BROWNLOW, C.B., 72ND HIGHLANDERS.

men in the heat of the action. He was a brave officer, well beloved by his Highlanders, and had led them to victory again and again—at the Peiwar Kotal, in the fighting around Cabul, and in the attack and capture of the Asmai heights. Captain St John Frome, of the 72nd, was also killed, and Lance-Sergeant Cameron, whom General Roberts in his despatch described as "a grand specimen of a Highland soldier." Captain C. S. Murray and Lieutenant Munroe, of the same regiment, were wounded ; as were also Lieutenants

Stuart, Menzies, and Donald Stewart, of the 92nd. Menzies
had a narrow escape. When capturing a walled enclosure he
suddenly found himself in an ambush of over 300 Ghazees,
whose leader, a tall and powerful fanatic, rushed at him with
a terrific yell, wildly brandishing his tulwar. Accepting
the challenge, the Highland officer sprang forward to meet
him. The trial of strength and skill lasted but for an
instant. Ere the Ghazee could deliver his first furious blow
the Highlander, with lightning rapidity, ran him through the
heart with his Andrea Ferrara claymore. Before, however,
he could extricate his weapon he was attacked by two of the
enemy from behind. They were in turn tackled and
despatched by a Highland corporal. Menzies was then
carried off and placed in a house, but was no sooner left alone
than he was stabbed in the shoulder by a Ghazee, who crept
in after him through an open window. This adventurous
fanatic immediately fell a victim to the kookerie of a gallant
little Ghoorka who saw the act.

After a time of most severe fighting, during which most of
the men of both Highland regiments performed prodigies of
valour, the two brigades emerged "at the point of the hill
near Pir Paimal, and bringing their left shoulders forward,
they pressed on and swept the enemy through the closely
wooded gardens and orchards which cover the western slope of
the hill." Some time afterwards a portion of the enemy
retired to a strong position—an entrenched camp—on the
Baba Wali Kotal, and to this they pushed forward reinforcements with all haste, determined to resolutely defend it.
This position it became necessary to carry at once by storm,
and Major G. Stewart White immediately called out to the
advanced companies of the 92nd he was leading—"Just one
charge more to settle this business."

"Joyfully and with alacrity," says Grant, "the Highlanders responded to the call of their favourite leader, and, without pausing to recover breath, drove the enemy from their entrenchments at the point of the bayonet." Regarding this movement Roberts said—"Nothing could be finer than the rush made by those two regiments, the Ghoorkas and the Highlanders." "The gallant Stewart White," continues James Grant, "ever foremost, was the first to reach the enemy's guns, being followed by the Sepoy Inderhir-Lama, who, placing his rifle upon one of the guns, exclaimed that it was captured in the name of the Prince of Wales' Own Ghoorkas. Another was secured by Major White, and special mention was made of this when he received the Victoria Cross. Here ensued, perhaps, the heaviest fighting of the day."

When it was over it was found that the Afghan Force was quite defeated, and Ayoub Khan's camp was at the mercy of the relievers. Thus was the battle won, Candahar relieved, and the object of the famous march accomplished. Among those who had distinguished themselves by "special gallantry and forwardness" General Roberts noted the following belonging to the 92nd Highlanders :—Major G. Stewart White, Lieutenant C. W. H. Douglas, Corporal William M'Gillivray, and Privates Peter Grieve, John M'Intosh, and D. Gray. Sentinels of the 92nd were placed to guard Ayoub's tent, which was most luxuriously furnished.

This battle practically ended the warfare for Sir F. Roberts and the brave Highlanders. Both had acquitted themselves with distinction, and on both were bestowed honours and rewards by a grateful Sovereign and admiring country. Before proceeding to India the last act of General Roberts was to distribute distinguished service medals to the Highland regiments and to the 5th Ghoorkas. The General addressed the troops in the following terms :—

"Soldiers of the Candahar Field Force,—I am glad to have

this opportunity of giving medals for distinguished conduct to the men of the 72nd and 92nd Highlanders and the 5th Ghoorkas. They have deservedly won them. I say from my experience as a soldier that no men with whom I have served could have better deserved these rewards, and it is an additional pleasure to me to have seen the other day of what material my Highlanders and Ghoorkas are made. I can but hope it may be my good fortune to have such good soldiers by my side when next I go into action. The 72nd have, I grieve to say, to mourn the loss of their Colonel, as fine a leader of men as I have ever seen; and with him fell an equally gallant spirit, Captain Frome, and many brave men, among whom I must mention Sergeant William Cameron, that grand specimen of a Highland soldier! But the 92nd had also a heavy loss, Colour-Sergeant Richard Fraser and other good soldiers being amongst the slain. On the 2nd September no less than fourteen gallant fellows were laid in one grave, and many of their comrades are now lying wounded in our hospital. But in all this you have a British soldier's consolation : that of knowing that you did your duty nobly. I believe in my day I have seen some hard knocks given and received, but never do I remember noticing a greater look of determination to win a battle than I observed in your faces on that morning of the 1st September! Not even the boldest Afghans could stand against such a bold attack. Yes! You beat them at Cabul, and you have beaten them at Candahar; and now as you are about leaving the country, you may be assured that the very last troops the Afghans ever want to meet in the field are Scottish Highlanders and Ghoorkas. You have indeed made a name for yourselves in this country, and, as you will not be forgotten in Afghanistan, so, you may rest assured, you will never be forgotten by me."

Chapter XXVII.

THE 92ND ON MAJUBA HILL.

IN the Transvaal war the interest for our readers centres in one event of most saddening disaster. The Boers had refused to submit to the annexation of the Transvaal territory by Sir Theophilus Shepstone to Great Britain in 1877. They protested and agitated for the return of their independence without producing any effect, and finally, on the 20th December 1880, they threw down the gage of battle, determined, if not to drive the invaders from the country, at least to make such a demonstration of their desire for freedom as to arrest the attention of the British Government.

The Boers, like others of the new enemies our troops have had to meet during the past decade, proved more difficult to beat than the military authorities calculated. They were athletic, well-trained men, knew the country, were admirable horsemen, most expert marksmen, and skilful in the strategy which could be employed with best effect in the kind of warfare in which they engaged.

The troops in the country at the time hostilities commenced were not nearly equal to the task of quelling the rising of the Boers, and among the reinforcements hurried forward to the assistance of those in straits was a number of the gallant 92nd Highlanders with their Cabul and Candahar honours fresh upon them. The 3rd of February 1881 found them in General Sir George Pomeroy Colley's camp at Prospect Hill. This was after the disastrous engagement at Lang's Nek, in which the deadly nature of the Boer fire had been terribly conspicuous, almost every shot telling with fatal effect.

The result of a reconnaissance of the Boer position by Sir George Colley was the desperate enterprise we are about to describe. Sir George determined to turn the flank of the enemy, and was confident this could be done by taking possession of Majuba Hill, a towering and difficult height on the left of Lang's Nek, at the base of which rested the right of the enemy's position. The hill was used as a look-out post by the Boers.

The critical movement took place on the evening of the 26th of February. On that night were detailed for service on a secret expedition 180 Gordon Highlanders, 150 of the 58th Regiment, 150 of the 2-60th Rifles, and 65 Blue-jackets, under Commander Romilly. In dead silence these men paraded near the General's tent. The "last post" had been sounded by the bugles, and all was quiet as they filled their ammunition pouches with the 75 rounds handed out to them. It was only when the little body marched off that its destination became known. At ten o'clock the word was given, and the men silently left the camp—"marching in fours, with their rifles at the trail." Three days' rations were carried. The General accompanied the Force, which was attended by several war correspondents, who shortly after shared in the hazards of the struggle which ensued.

"The night," says James Grant, "was pitchy dark at first, and the march across a country unknown to the men was toilsome in the extreme. At first the way was over comparatively level ground, but it was at the base of the hill the real difficulties began. . . . The path narrowed so much that after a time the sections of fours were diminished to Indian file, necessitating a sad delay ere the summit could be attained.

" At a precipitous part of the hill a company of Rifles was
left, and at the base one of the Highlanders. In their care
the horses were left. These men were all ordered to set
about entrenching themselves at once, while the remainder,
just as day was nearly breaking, and they were already worn
with a rough march of six hours, guided by Kaffirs, began the
ascent, a work of terrible toil, as in many places the ground
was most precipitous, the men having to crawl on their hands
and knees up dongas and over boulders, dragging their rifles
after them, up ways that even mountaineers might have
shrunk from in open daylight."

At length the summit was reached; the little British Force
occupied the Spitzkop, and had the whole Boer position in view.
Had this force been strong enough, indeed, the Boers were at
the mercy of the invaders. But, as it proved, this was where
Sir George Colley had erred. He had not enough troops
with him to consummate the great work contemplated; and
he had no guns or rockets with which to drive the enemy
from their position. Yet, well entrenched on such a post, the
troops at his command, if properly managed, should have
been able to keep any number of men at bay.

As soon as the plateau had been gained the General
ordered the troops remaining below to come up. He had
been successful in the first part of his rash expedition, but he
looked careworn and nervous, as if suffering from suppressed
excitement. What now happened may be learned from a
telegraphic description by the late Mr Cameron, of the
Standard:—" The enemy's principal laager was about two
thousand yards away. At sunrise the Boers were to be seen
moving in their lines, but it was not until nearly an hour
later that a party of mounted videttes were seen trotting out
towards the hill, upon which they evidently intended to take
their stand. As they approached our outlying pickets fired

upon them, and our presence was for the first time discovered. The sound of our guns was heard at the Dutch laager, and the whole scene changed as if by magic. In place of a few scattered figures, there appeared on the scene swarms of men rushing hither and thither. Some rushed to their horses, others to the waggons and the work of inspanning the oxen, and preparing for an instant retreat began at once. When the first panic abated it could be seen that some person in authority had taken the command. The greater portion of the Boers began to move forward with the evident intention of attacking us ; but the work of preparing for a retreat in case of necessity still went on, and continued until all the waggons were inspanned and ready to move away. Some, indeed, began at once to withdraw. At about seven o'clock the Boers opened fire, and the bullets whistled thickly over the plateau. The men were all perfectly cool and confident, and I do not think that the possibility of the position being carried by storm occurred to any one. From seven to eleven the Boers, lying all round the hill, maintained a constant fire. Their shooting was wonderfully accurate. The stones behind which our men in the front line were lying were hit by almost every shot."

While this long distance duel was going on the soldiers on the top of the hill blazed away a good deal of ammunition that could have been saved, and which would have been invaluable at the later stages of the conflict. Up to eleven o'clock there were but five casualties in the British ranks, but about this time, in full view of all, and while standing near the General, Commander Romilly fell mortally wounded. The first touch of the enemy at really close quarters was felt by a detachment of 20 Gordon Highlanders, under command of Lieutenant Ian Hamilton. They were stationed at a part of the hill under which the Boers gathered in strength and

suddenly opened a furious fire.   Hamilton was offered twenty
more men, but would accept only ten, and with these he
gallantly repelled the enemy's advance.   The men behaved
with splendid coolness, firing only when a Boer head was to
be seen.   Eight or ten Boers were believed to have fallen at
this point, but it is doubtful if the Highlanders' fire was so
effective as supposed.

    "So far," continues Mr Cameron, who was behind the
fighting line on the top of the hill, "our position appeared
perfectly safe.   The Boers had indeed got between us and the
camp, but we had three days' provisions, and could hold out
until the reinforcements came up.   From eleven to twelve the
enemy's fire continued as hot but as harmless as before;
between twelve and one it slackened, and it seemed as if the
Boers were drawing off.   This, however, was not the case.
Shortly after one o'clock a terrific fire suddenly broke forth
from the right lower slopes of the hill—the side on which the
firing had all along been heaviest.   A tremendous rush was
simultaneously made by the enemy.   Our advanced line was
at once nearly all shot or driven back upon our main position.
This position may be described as an oblong basin on the top
of the hill.   It was about two hundred yards long by fifty
broad.   Our whole force now lined the rim of the basin, and
fixed bayonets to repel the assailants.   The Boers, with
shouts of triumph, swarmed up the sides of the hill, and made
several desperate attempts to carry the position with a rush.
Each time, however, they were driven back with the bayonet.
After each charge the firing, which nearly ceased during the
mêlée, broke out with renewed violence, and the air above us
seemed alive with bullets."

    In the height of the terrible conflict the valour of the
Gordon Highlanders became conspicuous.   Their blood was
on fire, the fierce aspect and flaming eyes of the Boers rousing

them to a pitch of utter fury. They were exposed to a desperate fusilade; some of the slender force under Hamilton indeed recoiled—a movement which, rousing the wrath of gallant Ian Macdonald, caused him to draw his revolver and threaten to shoot dead the first man that passed him in flight. But a band of 150 brave fellows, most of whom were Highlanders, lining the ridge, stood stubbornly together. Thrice the Boers hurled their strength against them, and thrice they were driven back. In the intervals of the firing the men encouraged each other as they stood shoulder to shoulder exposed to the onslaught. "Don't budge from this; give them the steel here when they come," cried some, while the officers shouted words of encouragement. But the brave fellows were going down every instant. Here fell gallant Colour-Sergeant Fraser, of the Gordons, and many another who had for long years, and in not a few battles, upheld the honour of the grand old regiment.

But hard fighting was of no avail. The devoted band was outnumbered and over-matched. James Grant thus graphically describes the final stand :—"The whole Boer fire was turned on the last point of defence in the left rear. There the men were crowded behind a clump of stones, but the officers made them extend to the right and left, lest they should be outflanked. Our direct rear at one part was held by only thirty men; luckily the ground there was so steep the Boers were unable to scale it, thus all their efforts were hurled against the left. 'Men of the 92nd Highlanders, don't forget your bayonets !' cried Major Fraser. Colonel Stewart called on the men of the 58th, and Captain MacGregor on those of the Naval Brigade, and all did their duty steadily and well. In some places the Boers were seen, pipe in mouth, taking pot-shots quietly, as they do when practising at pumpkins rolling down a hill. Nearer and nearer the fatal cordon of death was closing round the devoted band on the hill of Majuba, and through the smoke the officers were seen doing their utmost to urge the defence. In the centre of a group that held a knoll was seen Sir George Colley, animating the men and behaving in the most resolute manner, though, one by one, they quickly dropped around him. With fixed

GENERAL SIR G. P. COLLEY.

bayonets, and shoulder to shoulder, at last, formed in semi-circle, our men continued firing, while ammunition began to fail. Many more fell, but there was no shelter to which they could be removed, and if there had been, not a man could have been spared to succour them. The Boers at last reached the men who held the true front; the latter brought their bayonets to the charge, but beyond striking distance, and all save three were shot down where they stood. With the General there were barely 100 men of the main body left. The advanced line had been long since shot down or driven in upon the last or main position. All at once Sir George Colley was seen to throw his arms above his head, to reel wildly forward, and fall dead, shot throught the brain, and then all was lost."

Even yet, however, the Highlanders who survived were not conquered. Their ammunition was done ; many of the foe could not be reached with the bayonet ; but as the Gordons sullenly fell back before that fearful, deadly fire they picked up great stones and fiercely hurled them at the coming foe. But one more rush and the struggle was at an end. The position was lost, and its brave defenders were massacred or in flight.

The loss of the British was great, the latest returns giving 85 killed, 131 wounded, and 60 prisoners. The Boer loss was trifling. It was the most terrible reverse suffered since Isandhlwana ; and it added to the severity of the disaster that the gallant General who led the troops had fallen.

The incident is one that, view it leniently as we may, fills us with concern. As a strategic scheme it was apparently as immature as it was daring. There was no cannon ; and the force wanted fighting coherence, being drawn from several instead of belonging to one corps. Then full advantage was not taken of the cover the hillside afforded to check the upcoming Boers ; much of the ammunition was blown away at nothing, and from positions where no one could be hit ; and, lastly, the General allowed some precious time to be spent in sleep on the hilltop, instead of utilising it in the immediate erection of entrenchments, which would have provided shelter for the men, and checked the enemy's rush. One thing alone there is no cause to regret—the valour of the soldiers, officers and men, and especially of the gallant

Gordons, who fought with all the determination, if not with the success, of their forefathers at Quatre Bras.

GRANITE MONUMENT TO GORDON HIGHLANDERS *(in Duthie Park, Aberdeen).*

Chapter XXVIII.

THE HIGHLAND BRIGADE AT TEL-EL-KEBIR.

THE chief interest of the Egyptian campaign of 1882 lies in the decisive action at Tel-el-Kebir. Here the trained soldiers of Britain met the trained Egyptians, and put them utterly to the rout. All the advantages of numbers and position lay with the soldiers of Arabi, but the domination of race asserted itself; the unconquerable determination of the British to win carried them over every obstacle.

It was a brilliant array of Britain's strength that marched over the sandy waste towards Tel-el-Kebir, where 30,000 Egyptians awaited the attack. Grenadiers and Coldstreams, with other distinguished regiments of the line, Household Cavalry, Lancers, Batteries of Artillery, a strong force of Engineers, dusky warriors from India, mounted and on foot —the flower, indeed, of the Imperial army, 14,000 strong. And what is of chief consequence to us, a genuine Highland Brigade was included in the force. Once more, as at the Alma, the 42nd—Duncan Macpherson of Cluny commanding —marched side by side with the 79th, and to these were added the 75th (the newly united 1st Battalion of the Gordon Highlanders) and the gallant 74th, a regiment with a proud record of brilliant deeds. It had fought under Wellington at Assaye shoulder to shoulder with the 79th, suffering terrible loss. It had gone through the long series of Peninsular battles, and its old flags bear inscriptions which tell of the glorious victories it had helped to win. For the conspicuous heroism of the regiment at Assaye leave was

U

given it to carry three colours—a distinction, we believe, permitted to but one other regiment in the British army. It had worthily sustained its honours in the Kaffir war, and had contributed its quota to the brave of undying memory who stood with Colonel Seton—erect, calm, and obedient to discipline—on the deck of the ill-fated *Birkenhead* while she sank into the depths of the Indian Ocean. The Highland Brigade formed part of the 2nd Division, which was under the command of Lieutenant-General Sir E. B. Hamley, and the Brigadier was one whom the 42nd Highlanders at least knew well, and of whom they were proud—Major-General Sir Archibald Alison. He had served with the 72nd in the trenches of Sebastopol; the friend of Sir Colin Campbell and attached to his staff, he had watched the 93rd at Lucknow hurl itself against the Secundra-Bagh and the Shah Nujjif, and although severely wounded, had written in terms of glowing eloquence of the deeds done that day; he had stood by Sir Colin's deathbed and received from the dying warrior the sword of honour which his father—the distinguished historian of Europe—in the name of the people of Glasgow had presented to the famous leader of the Highland Brigade; and he had led the 42nd to victory through the dense jungle in the famous rush on Coomassie.

At length the British troops neared the enemy's position. Arabi had tried to stem the advancing tide at Kassassin, but Sir Gerald Graham had hurled him back from his front after hours of hard fighting, during which it was found how stubbornly the Egyptians could do battle. Drury Lowe completed their route by a fierce charge of his heavy Cavalry Brigade. On the night of the 12th of September the British army was five miles distant from Tel-el-Kebir. Then Sir Garnet Wolseley gave the crucial order. The army was at 1 A.M. to march in the darkness of the night right up to the

trenches of the enemy without firing a shot, then carry them
by storm the moment they were reached, which it was calculated would be before daybreak. It was a daring undertaking,
this marching of a large body of men across the trackless

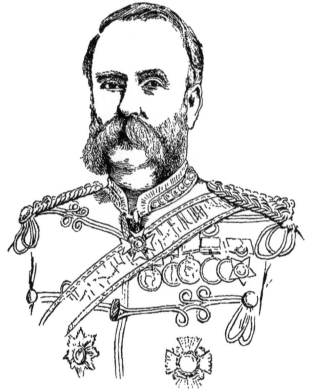

LIEUTENANT-GENERAL SIR ARCHIBALD ALISON, K.C.B.

desert, with nothing to guide them but the stars of heaven.
" Never," says Mr Cameron, of the *Standard*, " did a body

of 14,000 men get under arms more quietly. The very orders appeared to be given in lowered tones, and almost noiselessly the dark column moved off, their footfalls being deadened by the sand." The subsequent march and movements of the Highland Brigade have been eloquently chronicled by Sir Archibald Alison, who told the story of Tel-el-Kebir at a public gathering in Glasgow, where he received a sword of honour, and by Lieutenant General Hamley, who is known as wielding a most accomplished pen. Both of these distinguished soldiers speak chiefly of the conduct of the troops under their own command, and from them, responsible officers, and eye-witnesses we shall now largely quote :—
"Our instructions, generally stated," writes General Hamley in a paper contributed to the *Nineteenth Century*, " were to march on till we should come upon the enemy, and then to attack with the bayonet; and, the better to ensure the execution of this, I had directed that the Highland Brigade should not load. There was no moon, and the night would have been pitch-dark but for the stars. It was decided, after consideration of all contingencies, that my advance should begin shortly after half-past one. Accordingly, a little before that time I gave instructions to Colonel Ashburnham to move forward at a particular moment, and then rode forward to the Highland Brigade, which, though it extended half a mile across the desert, was not easy to find in the darkness. About half-past one it was called to arms, and about ten minutes afterwards the march on the enemy began. And here I must ask the reader to note that the northern half only of the enemy's line was the object of attack. There were no troops between my Division and the canal. For some distance onward the Engineers had erected a line of telegraph poles to guide us, but after they ceased the desert was absolutely trackless." Says Sir Archibald Alison—"The Brigade formed

for the march in the order in which it was to attack—two lines,
two deep. The rifles were unloaded, the bayonets unfixed,
and the men warned that only two signals would be given—
a word to 'fix bayonets,' a bugle sound 'to storm.' I never
felt anything so solemn as that night march, nor do I believe
that any one who was in it will ever forget it. No light but
a faint star; no sound but the slow measured tread of the
men on the desert sand. Just as the first tinge of light
appeared in the east a few rifle shots fired out of the darkness
showed that the enemy's outposts were reached. The sharp
click of the bayonets then answered the word 'to fix'—a few
minutes more of deep silence, and then a blaze of musketry
flashed across our front, and passed far away to each flank, by
the light of which we saw the swarthy faces of the Egyptians,
surmounted by their red tarbooshes, lining the dark rampart
before us. I never felt such a relief in my life. I knew
then that Wolseley's star was bright, that the dangerous zone
of fire had been passed in the darkness, that all had come
now to depend on a hand-to-hand struggle."

Then the British bugles rang out, and with lusty cheers
the Highlanders broke into the charge. "Without a
moment's pause or hesitation," writes General Hamley,
"the ranks sprang forward in steady array. Their distance from the blazing line of entrenchment was judged
to be about a hundred and fifty yards. In that interval
nearly two hundred men went down, the 74th on the left
losing five officers and sixty men before it got to the ditch.
This obstacle was six feet wide and four deep, and beyond
was a parapet of four feet high."

"On the right of the Brigade," continues Alison, "the
advance of the Black Watch was arrested, in order to detach
some companies against a strong redoubt, the Artillery from
which was now in the breaking light playing heavily on

General Graham's Brigade and our own advancing guns. So earnest were the Egyptian gunners here that they were actually bayoneted after the redoubt had been entered from the rear whilst still working their pieces. Thus it came about that, from both the flank battalions of the Brigade being delayed, the charge straight to their front of the Gordon and Cameron Highlanders in the centre caused these to become the apex of a wedge thrust into the enemy's line. The advance of these battalions was stoutly opposed by the Egyptians of the 1st or Guard Regiment, who fell back sullenly before them, and our men also suffered heavily from a severe flank fire from an inner line of works. Here one of those checks occurred to which troops are always liable in a stiff fight, and a small portion of our line, reeling beneath the flank fire, for a moment fell back. It was then a goodly sight to see how nobly Sir Edward Hamley, my division leader, threw himself amongst the men, and amidst a very storm of shot led them back to the front. Here, too, I must do justice to the Egyptian soldiers. I never saw men fight more steadily. Retiring up a line of works which we had taken in flank, they rallied at every re-entering angle, at every battery, at every redoubt, and renewed the fight. Four or five times we had to close upon them with the bayonet, and I saw these men fighting hard when their officers were flying. At this time it was a noble sight to see the Gordon and Cameron Highlanders now mingled together in the confusion of the fight, their young officers leading with waving swords, their pipes screaming, and that proud smile on the lips and that bright gleam in the eyes of the men which you see only in the hour of successful battle. At length the summit of the gentle slope we were ascending was reached, and we looked down upon the camp of Arabi lying defenceless before us."

With the characteristic modesty of a true soldier, General Alison notices the heroism of another, but takes no thought of his own.   While he dwells with pleasure on the "goodly sight" of General Hamley leading the men to the front in the thick of the fight, it is left to General Hamley to tell how his Brigadier behaved, and how in the front line, along with the Colonel of the 79th, was Sir Archibald Alison on foot. Brave leaders these of brave men—fit soldiers to lead to victory the Highland Brigade!

The victory of Tel-el-Kebir was gained in a short half-hour. The battle was fought and won ere yet the dawning morning had broken into daylight, and the flash of the hot fire maintained during its continuance was one of the means by which what was going on could be seen.   It was thus a difficult matter to correctly apportion praise or to single out men or corps who behaved with conspicuous distinction.   But in this instance the death roll tells too eloquently who had the stiffest of the fight.   The total British loss was 54 killed and 340 wounded.   The losses of the 2nd Division, of which the Highland Brigade was the front or fighting line, were 258 killed and wounded, of whom 23 were officers.   And, as General Hamley says, "any one interested in the question can, by referring to the lists of casualties, and comparing these with the losses of other bodies of troops, ascertain by that simple test on whom fell the brunt of the fighting."   It fell on the Highland Brigade.   Behind their point of attack the dead lay thickest, and the position presented the greatest difficulties.   Honour has already been done to the memory of young Donald Cameron (of the Cameron Highlanders), the Carse braes ploughman, who was, in the words of the *Standard* correspondent, "the first man to mount the trenches and the second man to fall."   The period at which this young hero was slain was one that told heavily on the Highlanders.

The Egyptians were fighting with the fury of desperation ; the trenches were deep and steep, and the climbing mass of officers and men were a ready mark for Arabi's sharpshooters. Major Colville and Lieutenant Somerville, of the 74th, went down, with young Lieutenants M'Neill and Graham-Stirling, of the Black Watch ; and Captain Hutton, the Brigadier's A.D.C. ; and Blackburn, Malcolm, and M'Dougall, of the Cameron Highlanders, were wounded. Captain Coveny—poor Coveny, since killed at Kirbekan —was also wounded, along with Captains Brophy—since drowned in the Nile—Fox, and Cumberland, of the Black Watch, and Midwood and Casey, of the 74th.* Sergeant-Major M'Neill, of the Black Watch, one of the most popular and estimable men in the regiment, and a splendid soldier, also fell, pierced by three bullets, after his claymore had sent six of the enemy to their account. And many a good man of the rank and file was struck out of the muster roll.   " Yet," wrote Cameron, of the *Standard,* " it was after these outer trenches were carried that the greater part of the casualties occurred. A few feet only in front of one of the bastions six men of the 74th were lying, heads and bayonets pointed forward ; while just in front of them, shot through the head, was the body of young Lieutenant Hume Somerville, who was evidently leading them on when a volley laid them all low." But the loss of the enemy was greater. " At some of the bastions," says the same correspondent, " the resistance, although unavailing, was desperate, the Egyptians being caught as in a trap by the rapidity of our advance, and defending themselves to the last. At these points the enemy lie dead in hundreds, while only here and there a Highlander lies stretched among them, lying face downwards, as if shot in the act of charging. In several places I counted from

* Lieut. H. G. Brooks, 1st Gordon Highlanders, was also killed, and Lieut. A. Graham Pirie mortally wounded in the engagement.

thirty to fifty lying in heaps, and they lay in rows where the 42nd, getting in flank, enfiladed the lines they were holding against an attack in front."

When resistance seemed at an end, the officers called out for a short halt, as necessary after so rapid an advance. "But," says Hamley, "just then Arabi's camp, occupying all the flat ground between us and the canal, was visible just below us, the last occupants escaping as a shell or two from the guns burst near; a body of his Cavalry was also forming near the tents, some of its officers riding forward as if to lead a charge upon us; and, pointing to these, I called on the men to make another effort and complete their work. They responded cheerily, and went right through the camp, capturing all the tents standing, with immense quantities of forage and provisions, and herds of loaded camels; while the Cavalry, its movement probably accelerated by a well-pitched shell or two, turned and galloped off." In the other parts of the field the same success had attended the British arms, and now the Cavalry, under Drury Lowe, was dashing over the country cutting and slashing at the fleeing fugitives. The splendid fighting qualities displayed and admirable discipline maintained by the Highlanders excited the admiration of General Hamley. In the *Nineteenth Century* article, already quoted from, he says :—" The Scottish people may be satisfied with the bearing of those who represented them in the land of the Pharaohs. No doubt any very good troops, feeling that they were willing, would have accomplished the final advance; but what appeared to me exceptional are—firstly, the order and discipline which marked that march by night through the desert; and, secondly, the readiness with which the men sprang forward to storm the works. The influence of the march had been altogether of a depressing kind—the dead silence, the deep gloom, the funereal pace, the unknown obstacles and enemy. They did not know what was in front, but neither did they stop to consider. There was not the slightest sign that the enemy was surprised—none of the clamour, shouts, or random firing which would have attended a sudden call to arms. Even very good troops at the end of that march might have paused when suddenly greeted by that burst of fire, and none but exceptionally good ones could have accomplished the feats I have mentioned."

The capture of Tel-el-Kebir was no mean feat in modern warfare. The numbers engaged brought it within the category of important battles; the numbers killed proved how fiercely the battle had been contested.

The performances of the Seaforth Highlanders (72nd) attached to General Herbert Macpherson's Brigade have not fallen within the scope of our narrative of the work of the Highland Brigade. Yet the 72nd did good work in the campaign. They were the hardened heroes who had done the march from Cabul to Candahar, and had little more than landed from the East when 200 of them advanced and engaged the enemy at Chalouffe. Their dash and coolness carried the day; the place was captured, the enemy was routed, leaving over 100 dead and wounded, and the waters of the Sweet Water Canal were allowed to flow. The regiment continued its advance with the Indian contingent, performing its full share of the hard toil and harassing duties. In the final dash on the enemy at Tel-el-Kebir it charged with all the fiery enthusiasm which had marked the conduct of the "Albany's" at the Peiwar Kotal, Cabul, and Candahar; and, ably led by Macpherson, was soon swiftly following the fleeing rebels.

Chapter XXIX.

THE HIGHLAND REGIMENTS IN THE SOUDAN.

THE two Soudan campaigns have given further opportunities of obtaining distinction ; but these events are so recent that they require to be little more than mentioned. When the square, under Sir Gerald Graham, on the 29th February 1884, marched out of Trinkitat towards Tokar, it was composed of men of a different calibre from those who had, under Baker Pacha, broken and fled before the Arab rush a few weeks previously. In front was the 1st Battalion of the Gordon Highlanders, the old 75th, and bringing up the rear of the square were the stately ranks of the famed Black Watch. On the right of the square were the Royal Irish Fusiliers, and on the left the York and Lancaster Regiment, while Rifles, Marines, and Engineers filled up the centre, and guns occupied the corners. At eight o'clock in the morning of the 29th of February the square marched out, the line of march being the same as that followed by Baker's men in their disgraceful flight, and one calculated to shake the nerves of young soldiers, for the way was strewn with the skeletons of the killed, bits of flesh and scraps of clothing still sticking to the bleaching bones.

Shortly after eleven o'clock the enemy was found at the wells of El Teb, and the square moved to the left flank of the position, then made a turning movement, for the time being throwing the Black Watch to the front. Straight at the enemy went the square, the Black Watch leading with all its old dash against the Arab stronghold, which included entrenchments, rifle pits, and two old forts armed with Krupps and Gatlings.   For an hour the enemy had been keeping up a fairly well directed fire, and as the square drew closer to the wells the resistance of the Arabs grew more fierce and desperate, and men were dropping every few steps. The wild Soudanese rushed fiercely on the square, and lunged their long sharp spears against the solid ranks of the Highlanders; some of them even found their way inside the square, and, hacking and slashing with all their might, were bayoneted where they fought.

The brunt of the Arab onset fell on the Black Watch and the Naval Brigade; but the swarthy warriors could not cope with the splendid discipline of the whitefaced soldiers, and had at length sullenly to retire.   When the main body of the rebels was seen drawing off in the direction of Tokar, the Cavalry were let loose, and dashed after them, inflicting sad havoc; but even in this conflict the Arabs maintained their courage, fighting fiercely and doggedly to the death.   The losses of the enemy were estimated at from 1,500 to 2,000, while that of the British Force was 31 killed and 142 wounded—the wounded including Baker Pacha and Colonel Burnaby.   The Arabs were thoroughly defeated, Tokar was occupied, and the force retired to Suakim, from which base it was intended to operate westward against the forces of Osman Digna.

The battle of Tamai which followed proved one of the most trying engagements in which a British Force had ever engaged. On the morning of the 13th of March 1884, the British troops marched out from Baker's zareba in double square, the Second Brigade, under General Davis, being composed of the 42nd Highlanders and 65th Regiment, guns being in front and the Marines in the rear.   This square advanced firing against the enemy, who were evidently in strong force in front, and occupying ground well suited to afford cover and concealment. At length the crowd of Arabs in front thickened, and they began to rush upon the square, the right flank of which was skirted by a deep nullah. The 42nd first half-battalion were ordered to charge, and as they did so straight ahead, and not where the enemy was strongest, a wild yelling suddenly arose from the nullah on the right, and hordes of Arabs were seen suddenly springing up from their concealment, and rushing furiously against the flank, now unfortunately opened by the charging movement in front. On crowded the Arabs, and the flank and front opened a furious fire to check their wild advance. But it was futile. Suddenly a confused roar rose above the din of conflict, and it took but a glance from onlookers to see that something was sadly wrong. The wild Hadendowas had reached the square, and were forcing it back. The open gap in the formation had been discovered by the enemy, and they had got inside, slashing and spearing with all their fury. The fighting grew desperate. British officers and men saw the disaster that was facing them, and strove their utmost to retrieve the day. But the Blue-jackets, driven back, lost their guns, and the front and right flank of the square were beginning to press on the rear ranks. Inside, every man, of whatever rank and profession, joined in the fight. There was no stampede, no rout. The 65th and the Highlanders were falling back, but they were fighting every inch of the way, and every man conquered or died with his face to his savage foe. A correspondent, who had been in Baker's ill-fated square at El Teb, thought he saw the same picture again presented to his eye, but he recognised the difference. Instead of faltering, cringing Egyptians fleeing before the naked howling Arabs, here were valiant British soldiers, fighting for honour and country, bearing themselves proudly in the dire conflict; and with their passions thoroughly roused and the battle fervour bounding in their veins, they buffeted with fist and foot, and struck with butt and steel the desperate fanatics, who crawled and dodged, leaped and yelled, and slashed in and around the square.

"The spectacle," wrote the correspondent of the *Daily News*, "did not so much terrify as exercise a weird, terrible fascination. I do not suppose that either I, or anyone else who witnessed it, will often again see its equal for magnificence. Though retreating, our men literally mowed down their assailants. In the smoke and dust of the battle, amid the bright gleam of their myriad spearheads, the semi-nude, brown-skinned, black, shaggy-haired warriors were falling down in scores. Of all the savage races of the world none are more desperately brave than the Soudan Arabs, who were breaking upon our ranks like a tempestuous sea. At last the pressure of the front upon the rear became so great that those of us who were mounted were for a few moments too tightly wedged together to be able to move; but we felt the collapse was only temporary." During the conflict many deeds of personal valour were performed. Bennet Burleigh, the intrepid war correspondent of the *Daily Telegraph*, mentioned some of these. "One private," he says, "ran at a furious Arab, and bayoneted him so violently that his rifle point actually entered the man's body, and the private dragged the wounded man along before he could withdraw the weapon. Of twenty men of the Black Watch who were in the first charge up to the nullah's edge only three survived the terrible conflict. These twenty did not fire their weapons, but used the cold steel and the butt-end of their muskets. The fighting was far too close and fierce in this *mélée* to allow the men withdrawing their cartridges and so keeping up a fire. Two powerful Highlanders, Jamie Adams and Donald Fraser, made a dozen foemen bite the dust before they themselves fell from loss of blood and the severity of their wounds. A horseman on a grey charger, who was ascertained to have been Osman Digna's cousin, Sheikh Mahomed, struck wildly at Private Drummond while that gallant soldier was bayoneting an Arab. Though stunned by the blow, Drummond rallied, and killed the Sheikh. As Drummond was engaged fighting the Sheik, another Arab rushed at him, only to be shot by Drummond's chum, Kelly, who himself was instantly killed."

At length the grand example of the officers and the heroic determination of the men checked the fearful onslaught. The tide of movement turned. Shoulder to shoulder the 42nd and 65th first stood firm, then advanced against the enemy, slowly at first, then quicker, as they felt the triumph of conquest rekindle within their breasts. Then their wild cheers rose on the air above the roar of musketry ; the crisis had passed ; the Hadendowas were beaten.

In that terrible few minutes the column had lost heavily— 109 killed and 111 wounded ; but 3,500 Arabs had sacrificed their lives. Of the British loss the Black Watch contributed 66. Two of its officers were among the wounded.

In this hotly contested fight the 1st Battalion Gordon Highlanders formed half of the front and the whole of the right face of the 1st Brigade square, commanded by Sir Redvers Buller. The losses of the Gordons and the other corps composing the 1st Brigade were happily not severe, which was no doubt due to the fact that the body of the enemy who attacked them had to pass over about 250 yards of moderately open ground, under a steadily aimed rifle fire, which the Highlanders delivered with great coolness. General Buller's Brigade in the crisis afforded most welcome assistance in distracting the attention of the enemy who were engaged in pressing back General Davis' Brigade.

There is no need to describe the latest engagement in which the 42nd have taken part. Little more than a year has elapsed since at Kirbekan they attacked the enemy with all their old dash and heroism. Strongly posted as the Arabs were, the Highlanders, along with the Staffordshire Regiment, boldly advanced against them, and with pipes playing and their cheers rising on the desert air, drove them from an almost invulnerable position. Here well-beloved Colonel Coveny fell, along with General Earle. The 1st Battalion Gordon Highlanders also took part in this harassing expedition. The Battalion was not lucky enough to arrive at Kirbekan in time for the action, but it accompanied the river column through all its arduous ascent of the Nile, to within twenty miles of Abu-Hamid, where the whole column received orders to turn back. In Lord Wolseley's Nile boat race the Battalion was the second in point of time in performing the passage from Sarras to Korti, following closely on the Royal Irish, who gained the prize.

Since then the brave Camerons have been engaged in a thankless task in the Soudan—forming at Koshay the solitary barrier between the Arab hordes and Egypt. They have fought firmly and well, perhaps earning less than their share of glory, but performing a very full share of trying and hazardous duty.

And now we have concluded our story of Highland valour as exhibited in the battles of our own time. We have not professed to find the Highland soldiers better or more valorous men than they were eighty or a hundred years ago; but we have seen no signs of degeneracy. Against foes differing widely in their numbers, discipline, tactics, and courage, they have been again and again pitted ; and, if we except Majuba Hill, with but one result—the wresting of victory from those against whom they fought.

APPENDIX.

A—PAGE 30.

THE language of this sentence, as the reviewer in *Vanity Fair* has pointed out, rather overstates the achievement of the Highland Brigade. In the advance, the Highland Regiments were accompanied by the battalions of Grenadier and Coldstream Guards attached to the First Division. These battalions, who entered the battle just before the Highlanders, were able to seize and hold the Redoubt, while the Highlanders stormed the heights further to the left. The battle had, however, notwithstanding the brilliant nature of the Guards' advance, not passed the critical stage until the Highlanders went forward. But the honours of victory were fairly shared by the Coldstreams and Grenadiers; while the Scots Fusiliers fought and suffered with desperate valour.

Another point on which *Vanity Fair* criticised me rather contemptuously, was my "fond acceptance" of what my reviewer called Kinglake's "picturesque inaccuracy," that the Highland Brigade advanced at the Alma wearing their feather bonnets. These, says the reviewer, were mostly returned to the regimental store on the memorable 20th of September. In his own Journal I tried to set him right, but he held to his view, backing it up with the following personal statement—"I heard Colonel Stirling, who accompanied Sir Colin during the latter's term of duty as Inspector-General of Infantry, speak of the matter in the presence and in the hearing of his Chief. He regretted that the Highlanders did not wear their plumes at the Alma. This was at the Mess of the Templemore Depôt Battalion, in the autumn of 1856, when the facts were still fresh in the memory of many of the officers present at table who had also been present at the battle." I confess that I had followed Kinglake's narrative in this particular, but I had never, in all I had heard of the engagement, known of the statement that the Highlanders wore the feather bonnets being questioned ; nor had I ever read of Kinglake, who was present on the field, being controverted on this point. Yet, another correspondent of *Vanity Fair* supported the critic in his averment. While the discussion was proceeding in the columns of that journal, I had no time to obtain testimony that would be considered indisputable. But I was far from satisfied, and have since made inquiries, with the result that my contention and the accuracy of Kinglake's description is fully established. (1) Col. Cooper, 4th Royal Irish, who carried the flag of the 93rd during the engagement, writes me—"The 93rd certainly wore the feather bonnet at the Alma." (2) General J. A. Ewart, who was present with the 93rd at the Alma, in his "Story of a Soldier's Life," says—"A Russian General, who was taken prisoner, stated that their infantry would not stand firm after they caught sight of the bare legs and *waving plumes* of the Highlanders" (vol. i., pp. 234, 5). I cite these statements merely as corroborative of my position ; but (3) the following letters which I have received from General Sir John Douglas, who commanded the 79th at the Alma, and

from General F. W. Traill-Burroughs, who was then Lieutenant in the
93rd, place the matter beyond a doubt, and I rest upon them as conclusive :—

"Glenfinart, Greenock, N.B., Dec. 6th, 1886.   Dear Sir,—I had
no doubt about the matter of the feather bonnet myself ; but to
make more certain, I wrote to Col. Ainslie, who commanded the 93rd,
and to Col. Pitcairn of the 42nd, and they both say that the Highland
Brigade wore the feather bonnet at the Alma, and during the whole
Crimean War.—Yours truly, JOHN DOUGLAS, General."

"Rousay, Orkney, N.B., 8th Dec. 1886.   Dear Sir,—I have no hesitation in saying that the 42nd, 79th, and 93rd all wore the feather bonnet
at the battle of the Alma.   The 42nd was the only regiment that wore
their knapsacks.   The 79th and 93rd carried their greatcoats rolled and
slung over a shoulder and round the body, and in the rolled coat carried
a change of linen, &c.   After the battle, the next day I think, the Duke
of Cambridge assembled the Division of Guards and Highlanders, and
thanked us for our conduct at the battle ; and Sir Colin Campbell told
his Highlanders that he had asked Lord Raglan as a favour, and out of
compliment to his Brigade, for himself and his Staff to be permitted to
wear the feather bonnet, the national head-dress of the 'people of
Scotland.'   There was much joking about the feather bonnet being
called the national head-dress of the people.—Faithfully yours, F.
BURROUGHS."

These letters I regard as incontrovertible proof, and I am, therefore,
driven to the conclusion that on the occasion referred to by my critic,
Colonel Stirling had been speaking of something else.

B—PAGE 34.

General Burroughs, writing regarding the expeditions to Kertch and
Yenikale, says :—"They were rather pic-nics than hardships, excepting
that we were very closely packed in the men-of-war and transports that
conveyed us.   The weather was delightful ; the Black Sea was calm,
and the scenery beautiful.   And the enemy did not dispute our dis-
embarkation, or come near enough to molest us.   Kamara, where the
Highland Division was hutted in the Crimea, was a lovely spot, and
we had less hardships there than at any time during the war."

In reference to a statement near the end of the first paragraph in
page 34, General Sir A. Alison writes :—"I was not Adjutant of the
72nd, but Captain of the Light Company, during the campaign in the
Crimea.   I had been for about two years Adjutant of the Depôt, but
was promoted to a Company shortly before the Regiment embarked for
the Crimea."

C—PAGE 41.

General Burroughs writes :—"The 'angry Scotch wife' is still, or
was lately, alive.   Her name is Mrs Smith.   She was afterwards known
by the Turks, encamped near us, as Kokand Schmid (Kokand is the
Turkish for woman).   Her husband was Quarter-Master Donald Sin-
clair's batman, and she was one of the few women permitted to accompany the Regiment to the Crimea.   She and her husband, I believe,
are both still alive, and living at Bell's Mills, Dean Bridge, Edinburgh.

They were a most respectable couple when I knew them, and I perfectly remember witnessing the event you relate."

D—PAGE 136.

General Sir Herbert Macpherson, V.C., K.C.B., K.C.S.I., died of fever on board a steamer at Prome, in Burmah, on October 20th of this year. In August preceding, while holding the office of Commander-in-Chief of the Army of Madras, he was instructed to proceed to Burmah and take command of the Forces there engaged operating against the Dacoits. He reached Mandalay on the 17th September. After a stay of some time in the capital, in a most unhealthy atmosphere caused by recent flooding of the town, he descended the Irrawaddy to Thayetmyo, where fever symptoms developed themselves. He continued his journey down the river, intending to reach Rangoon, but had only got as far as Prome when he died as stated. The death of General Macpherson is a loss, not only to the Indian but to the British Army. He was a brilliant soldier, and popular with all who served above or under him. He belonged to an old Nairnshire family, and was born at Ardersier, near Fort-George, in 1827. He was the son of Duncan Macpherson, who had raised parties of men to serve in the famous Ross-shire Buffs (the 78th Highlanders), and who had risen to command the Regiment. In the 78th young Herbert served his novitiate in the profession of arms. He was gazetted an ensign on the 26th February 1845, and was lieutenant in January 1848. He went with his regiment through the short Persian campaign, and was adjutant when it joined Havelock to engage in the terrible struggle to relieve Lucknow. He was twice wounded before the Residency was reached. It was on the day of the final advance, as will be found stated in the text, that he gained the distinction dearest to the soldier's heart—the Victoria Cross for conspicuous valour. In 1857 he became a captain, in 1858 a brevet-major. When the 78th left India for home, Macpherson joined the Bengal Staff Corps and remained. He was immediately appointed by Lord Clyde to the command of a Ghoorka regiment. In 1867 he was lieutenant-colonel; in 1872 he attained to the rank of full colonel. In the first advance into Afghanistan in 1878 he commanded a Brigade under Sir S. Brown, and distinguished himself by his indefatigable labours. A Companionship of the Bath was the distinction awarded for his services. His conduct in the subsequent campaign under Sir Frederick Roberts—when the 92nd Highlanders were brigaded under him—was of the most brilliant and dashing kind, and he was rewarded with a Knight Companionship of the Bath. Shortly after, he visited Inverness, and was the guest of his kinsman, Provost Macandrew. During this visit the Northern Capital honoured itself by conferring upon the distinguished soldier the freedom of the burgh. At Tel-el-Kebir he was General in command of the Indian contingent, and performed the work assigned him with a dash which won the admiration of his companions in arms, and the warm encomiums of the Commander-in-Chief. It was he who, after the battle, pushed on with his troops after the flying enemy to Zagazig, capturing five trains filled with armed men, then he went on to Cairo, and completed the triumph. The thanks of Parliament, a medal with clasp, the Khedive's star, the second class of the

Medjidieh, and being created a K.C.S.I., were the rewards he received. Again he visited Inverness, and a sword of honour was presented to him at a public banquet there. He returned to India to assume the command of the Allahabad Division of the Indian Army; from that he was promoted to the command of the Army of Madras—the position he held when he died as above described. Sir Herbert was, as a writer has well expressed it, "a genuine soldier, in all respects a man of deeds and not of words; without fortune and without influence, his was a career that forms a remarkable instance of what personal bravery and loyalty to duty may attain."

E—Page 167.

Thirty years have not abated the determination with which this claim is disputed. General Burroughs and Colonel Cooper have each favoured me with statements, in which they re-assert their title to the distinction of being the first to enter the breach. Colonel Cooper says :—

"It is not the fact that Captain Burroughs was the first to enter the breach of the Secundra-Bagh. No one saw him go in, nor was he voted for by the officers of the 93rd. As a matter of fact, he did not receive a single vote. Colonel Leith-Hay may have recommended him for the Victoria Cross, but Colonel Hay was not present at the storming of the Secundra-Bagh, having gone with the right wing of the regiment towards the Barracks. How could Brigadier Hope have supported Captain Burroughs' claim, as you say, when he did not vote for him? You have no doubt read the 'Records of the 93rd' and the letters testifying to my having been the first; besides, Captain Blunt, Bengal Artillery, saw me go in. I have written statements from General Ewart which differ very materially from what he wrote in 1880. Now, sir, may I ask whether you ever heard that Captain Burroughs ran back, shouting—'Go back, go back, all of you'? which was the cause of Colonel Ewart and me being so badly supported. It would appear from your book that Captain Burroughs was the only officer who served throughout the Crimean War. I beg, however, to state that I landed (in spite of the omission in General Ewart's book) with the 93rd, carried their colours throughout the war, and never left till the army embarked for England. I do not wish to say anything unkind or harsh about your book, but it appears to have been written for the benefit of Scotchmen, to the detriment of an Irishman. But the truth in history will outlive all else."

As my narrative of the attack and capture of the Secundra-Bagh was written before I had held any communication with General Burroughs, and remained unaltered by any suggestion of his, I thought it desirable to bring under his notice the main points of Colonel Cooper's statement, to afford him an opportunity, if he so desired, of giving his own version of the incident to place beside that of Colonel Cooper. He has written me as follows :—

"I can only reiterate all I have already said, and add that I am very sick and tired of this controversy, which has been going on for some thirty years, as to who was the first man through the breach made by the British guns in the Secundra-Bagh. Could I have known it would lead to all the bickering and correspondence it has done, I would not have led the way through the breach, but would gladly have allowed Lieutenant Cooper, had I seen him, or any one else to do it.

"Colonel Cooper can, of course, say what he pleases, but I take my solemn oath that no man entered the breach before I did. I was the first to arrive at it, as I have before described, and when I reached it it was not large enough to admit a man through, and it was not until I and the men with me, who saw me do it, and whose names are mentioned at p. 203 of Burgoyne's 'Records of the 93rd Sutherland Highlanders,' had pulled down the masonry around the hole in the wall made by our Artillery, that I was able to, and did, lead the way and scramble through it.

"I had no thought of the Victoria Cross when I started for the breach, and did not know that I had done anything extraordinary after it was done. I was favourably placed for crossing the intervening space of ground to the breach, and, hearing the order for the 93rd to advance, I shouted to my Company (No. 6) to follow me, and made straight for the breach. Anybody else would probably have done the same. I made no boast of what I had done, and it was not until I heard that Sir Colin had said that 'There never was a better feat of arms'; and when I heard that another was claiming the credit of being the first through the breach, that I went to Colonel Leith-Hay and to Brigadier Adrian Hope and told them what I had done, and gave them the names of the men who were with me and saw me do it.

"Although Colonel Cooper appears so anxious to obliterate all I did, I have no wish to detract in any way from his bravery on that occasion, excepting in contradicting his assertion that he was the first through the breach, which I most emphatically deny."

So far as the statements of the officers chiefly concerned go, the dispute is left precisely where it began so many years ago. With all respect to Colonel Cooper, and freely acknowledging his perfect sincerity in the matter, after full consideration of everything I can find in his favour, I still stand by the narrative in the text. The book was, I admit, written for Scotsmen—from the first line to the last—but not to the detriment of either Irish or English soldiers, many of whom, in the campaigns described, with invincible bravery upheld the honour of the Highland regimental flags. As a matter of fact, I found out only after I had written of the Secundra-Bagh affair that General Burroughs was a Scot. When writing, I was under the impression that he was an Irishman, and the correction was made in the proof slips. But that aside. In regard to the point in dispute—will Colonel Cooper pardon me pointing out to him that the regimental Victoria Cross voted for, and for which Captain Burroughs did not get a vote, was not bestowed for being first through the breach in the Secundra-Bagh at all, but for conspicuous valour displayed during the action? This is clear from the fact that Captain Stewart was awarded the Cross by the highest number of votes (18), and he was not near the breach, but gained the distinction for his conduct in another part of the field (see page 163). Then Lieutenant-Colonel Ewart got 16 votes, and we have his own testimony ("Story of a Soldier's Life," vol., ii. page 77) that he was not the first through the breach, although for capturing two flags as described he at one time hoped for the Victoria Cross as his reward. And Colonel

Cooper himself surely had other claims to the Victoria Cross than were represented in being first through, for all history of the action shows that no man that day excelled him in valour until he was cut down in the heat of the struggle. General Burroughs' action never was brought to the vote—the recommendation for the Victoria Cross concerned himself alone, and was for the specific act named. There is, therefore, nothing necessarily inconsistent in the statement that Adrian Hope supported Leith-Hay's recommendation. Two points more remain—(1) when General Burroughs reached the breach he could not get through until the hole was made big enough. Colonel Malleson, who espouses Colonel Cooper's cause in the controversy, represents that officer as "flying, so to speak, through the hole"; and he quotes Captain Blunt, of the Bengal Artillery, who says "his jump into it reminded me of the headlong leap which Harlequin in a pantomime makes through a shop window." Is not the inference from this that Cooper must have reached the breach after it had been opened out by Burroughs and his following? (2) Lance-Corporal John Dunlay was presented with the Victoria Cross—I quote the language of the official record—"for being the first man now surviving of the regiment who, on 16th of November 1857, entered one of the breaches of the Secundra-Bagh, with Captain Burroughs, whom he most gallantly supportd against superior numbers of the enemy." That seems conclusive. I have always thought that a statement from Dunlay might settle the point, and have tried to get at him, but without success. I had never heard or read of Captain Burroughs interfering with the advance, as described by Colonel Cooper.

F—Page 168.

In this line, General Alison has in previous editions been described as "Aide-de-Camp," instead of "Military Secretary" to the Commander-in-Chief. Sir Archibald writes:—"I was not Aide-de-Camp to Lord Clyde, but his Military Secretary. My younger brother, Lieutenant Frederick M. Alison, who had been one of Sir Colin's A.D.C.'s towards the end of the Crimean War (and who was also in the 72nd Highlanders), was one of the A.D.C.'s during the whole of the Indian Mutiny; and was wounded, but not seriously, a few minutes before I was, at the assault on the Shah Nujjif, and we lay together on our litters on the field during the whole of the ensuing night."

G—Page 233.

I think an extract from a letter I recently received from General Burroughs may fittingly be inserted here. He writes :—"What has most impressed itself upon my mind after the perusal of the noble deeds of self-devotion and self-sacrifice at the call of duty, which you have so graphically described, is—What has become of those who have survived them and may be still alive? Too many of them, I fear, now are paupers and wanderers begging for their bread ! Should this be so? Is it right that the soldiers of so rich and powerful a country should so end their days? It is, I know, often their own fault that they come to want; but the very generosity of disposition that induces them so nobly to sacrifice their lives (as you have so truly described) at the call of duty, also too often makes them careless in their worldly circumstances. Having been for years accustomed to have every provision for their existence and comfort considered and carried out for them, when thrown upon civil life and their own resources they are not able to take care of themselves. To remedy this, I would suggest that every soldier on his discharge should be awarded a pension, according to rank, for every campaign and battle he has taken part in—irrespective of, and in addition to, the pension for length of service. And I would further suggest that an asylum for old, and especially for war-worn, soldiers be established in each military district, on much the same principle as Chelsea Hospital, London, to which they could retire within reach of their friends and end their days in peace, instead of as at present serving as scarecrows to scare away young men from entering the military service of a country apparently so careless of the fate of those who have so faithfully served it."

I know one or two of the heroes of the Sarda River struggle, and to some extent the language of the gallant General applies to their case. They are broken-down men—exhausted and worn beyond their age, unfit for hard work, and with a pension of but 8d. or 9d. a-day. Their being pensioners precludes them, when out of employment, from receiving parochial relief; yet they are expected, if householders, to pay poor's rates; their being pensioners also precludes them from obtaining employment or aid from "unemployed committees" in times of trade depression. The proposals of General Burroughs are probably too generous ever to be carried out—but our war-worn soldiers should really be protected against extreme poverty in old age. "The evil is a very real one," writes Sir Archibald Alison in reference to General Burroughs' remarks, "and I get one or two letters every week from old soldiers suffering under it, but the remedy is hard to find. It could, of course, be done by giving all men who had served a pension; but it was to avoid this great expense of pensions resulting from long service which was one of the strongest reasons assigned for the introduction of the short service system." The short service system may ultimately bring about the extinction of the "scarecrows," but meantime there are hundreds of suffering Crimean and Indian heroes amongst us; can nothing be done for them? In some quarters pensions are still distributed with a liberal enough hand. Could not a Committee of influential army men signalise the Year of Jubilee by obtaining a grant for the poor survivors of these two great wars?

H—Page 236.

Colonel Duncan Macpherson, C.B., of Cluny, died on Sunday, October 23rd, at Cluny Castle, near Kingussie, Inverness-shire, at the early age of 53. He had but a few months before retired from the service, and taken up his residence in his ancestral home, to which he had succeeded on the death of his father—Cluny Macpherson, C.B., a fine old typical Highland chieftain. Colonel Macpherson had been all his life a soldier, and all through his soldiering career was connected with the famous Black Watch. He was born on the 9th of October 1833, and entered the 42nd as ensign before he had completed his nineteenth year. He was lieutenant in 1854, and was captain when the regiment went to India to engage in the suppression of the Indian Mutiny. He was present at the battle of Cawnpore, where the Black Watch joined the army of the Commander-in-Chief, Sir Colin Campbell, and was engaged in the subsequent pursuit of the Rebels, and the battle of the Kallee Nuddee. He went through the hazardous fighting at the siege and capture of Lucknow, and the incident narrated on page 196 the author had from his own lips. He accompanied the regiment through the campaign in Rohilcund, and was present at the engagements at Fort Rooyah, Allygunge, and Bareilly. He was promoted to his majority in 1865. In 1872 he, representing the regiment, unveiled the Black Watch memorial in Dunkeld Cathedral, and placed thereon the flags carried by the regiment through the Indian Mutiny War. And with the rank of major he commanded the Black Watch in the Ashantee campaign during the famous advance on Coomassie, and was thrice wounded during the fighting that preceded the capture of Amoaful. His conduct on this occasion drew forth the admiring comments of the Brigadier (Sir A. Alison), who in his report to the Commander-in-Chief wrote—"I have particularly to bring to your notice the excellent manner in which the attack was conducted by Major Cluny Macpherson, and the unflinching gallantry which he displayed. Though wounded in three places, he positively refused to quit the field until the village of Amoaful, which I had directed him to attack, was secured." In the final attack on Coomassie, Colonel M'Leod took command of the regiment. For his share in this campaign the deceased officer was rewarded with a Companionship of the Bath, and was promoted to a brevet lieutenant-colonelcy. In the Egyptian campaign of 1882, Duncan Macpherson, with the rank of full colonel, commanded the 42nd, and once more led that gallant regiment to victory at Tel-el-Kebir. Although the Highland Brigade did not on this occasion receive the credit they expected in the report of the Commander-in-Chief, the written and spoken language of General Hamley, their divisional commander, and of General Alison, their brigadier, testifies conclusively that their conduct merited the highest praise. It is said that Colonel Macpherson felt that justice was not done himself and his regiment in official despatches. Shortly after he resigned the command of the Black Watch, and was appointed to the command of the Perth Military District—a post which kept him in close touch with his old comrades. He held this position till after his father's death, relinquishing it early in the present year to take up the Chiefship of the Clan. But ill-health had already overtaken the war-worn soldier. Exposure, fatigue, the rigours of climate, and the effects of wounds, had broken down the robust constitution, and Duncan Macpherson went home—to die. He was a strict disciplinarian, but a popular officer, and deeply beloved by his men. He was in private life cheery, affable, and generous. In a word—he was a good soldier and a good friend.

ERRATUM.—Page 164, line 15 from bottom, for "trenches" read "breaches."

Lightning Source UK Ltd.
Milton Keynes UK
UKHW010755160620
365053UK00001B/78